CONTRACT YOURSELF AND $AVE

CONTRACT YOURSELF AND $AVE

The Complete Guide to Owner Contractor Construction

Gary F. Wiggins

To order additional copies of this book, contact:
Xlibris Corporation
1-888-795-4274
www.Xlibris.com
Orders@Xlibris.com
14579

CONTENTS

Dedication

To my wife Joan, who has been with me every step of the way. Without her this book would not have become a reality. She is my source of encouragement, support and inspiration and, in the practical aspects of life, is more an expert than I ever hope to be in the construction industry

PREFACE

The most often asked questions about owner/contractor construction are: Is it all it is touted up to be? Can real savings be realized? Is it difficult? Do I need experience? Can I really succeed as an owner/contractor? Is it for me? The answers, naturally, are based on the degree of commitment the individual is willing to make. With a solid commitment, however, the answers are obvious. Yes, owner contractor/construction is all it is touted up to be. Real savings can definitely be realized. The degree of difficulty is directly proportionate to your understanding of the processes involved (which are addressed in this book). Although very desirable, experience is not absolutely necessary. You can succeed as an owner/contractor if you know what tasks to perform, how and when to perform them and are prepared to see the project through to completion. At this point, the answer to the last question is maybe. Given a fair reading of the following chapters and understanding the pros and cons discussed, however, by the time you finish this book, you will be prepared to answer the last question with a definite yes or no. There will be no maybe!

Almost everyone is intrigued by the thought and resulting challenge of building his or her own house. I have noticed that those people will fall into one of four very general groups. Group one, which I refer to as the "thinkers," consists of the largest percentage of people. Thinkers have the desire to build and may even get excited by the thought of building, but understanding their limitations in construction knowledge or experience, and considering personal time constraints, go no further than the talking stage. Group two, the "fence sitters," have the desire and ambition to build and definitely would do it if they only had someone with construction knowledge to assist or advise them from time to time. Fence sitters may lean in either direction, but will ultimately decide to go further or not, based on their comfort level at the time the decision is made. Group three are the "hardheads." Hardheads are determined to build, regardless of the circumstances. Hardheads tend to go ahead with a project before they are ready, usually do not ask for or listen to advice, make critical mistakes and, sadly, either fail to complete the project or fail to meet their goals. Group four, referred to as the "doers," is composed of the smallest percentage of people. Doers take the time to learn the processes, are not afraid to ask questions, stay informed and devote the time required to keep the project running smooth. When doers take on a major construction project, all tasks are known in advance and performed in a logical manner. Needless to say, doers ultimately meet their goals.

Although owner/contractor construction can be performed by anyone, it is not for everyone! It will only produce results for those people willing to perform all tasks that are necessary to see the project through to completion. This book is designed and written to inform you of the benefits and pitfalls, pros and cons, costs and potential savings and the rights as well as liabilities of an individual property owner assuming the role of contractor. The information provided is intended to present a realistic and objective view of all segments of the construction process. My personal goal is to encourage individuals to think about and understand the entire construction process before starting a project. At no point should you be lured into starting a project solely by the potential monetary savings. If you are not fully convinced that you can complete a project, do not start it. After reading and fully understanding the information as I present it, you will have what is necessary to make an intelligent and informed decision as to whether or not you want to take on the tasks associated with owner/contractor construction.

If you are a fence sitter or a doer and your decision is to assume the role of contractor, the information contained herein will definitely help you save money. If you are a fence sitter or a thinker and the information in this book discourages you from taking on a responsibility that is larger than you anticipated, or prevents you from starting a project then realizing you cannot or are not willing to complete it, you will still save money. Either way you will benefit financially from the information you are about to read.

I cannot put enough emphasis on the importance of reading and digesting all chapters before attempting to start a construction project of any magnitude. Constructing a house takes quite a bit of advanced planning and continuous, personal involvement at all stages. Realize that up front. Read the book at least once then reread all of the chapters that will apply to your project as many times as necessary for you to get a solid understanding of the techniques and procedures you will need to employ, before you start. If you decide to start a construction project, be prepared to continue it to completion. During construction, keep the book close at hand and refer to it often.

CHAPTER ONE

SITE SELECTION

Although there are many factors to consider in purchasing property for a personal single-family residence, the vast majority of people select a site based solely on location. All too often, sites with prime locations require an excessive amount of preparation before improvements can begin. Location should naturally be high on your list of priorities for site selection, but be sure to consider all the factors listed in this chapter before purchasing.

SUBDIVISIONS—With very few exceptions, you will not be able to develop a home site without falling within the jurisdiction of a city, county, township or other local government. In most cases, the local governing body will have already established standards for further subdivision and development of communities and neighborhoods within its boundaries, known as subdivision regulations.

In creating a subdivision, the developer must adhere to the design standards of the subdivision regulations for the local governing body. By the same token, the developer can and, often does, impose design standards of its own on the individual parcels or lots within the subdivision. Those design standards are known as deed restrictions or covenants. In some cases the design standards may be contained in the property deed itself and in other cases in a separate accompanying document. In all cases, however, the restrictions should be reviewed and understood before purchasing your site.

Deed restrictions usually contain such vital information as height limitations on structures, minimum square footage allowable within structures, minimum distances structures must be located from property lines or accessory buildings and allowable use of the property.

EASEMENTS—Easements, which are usually not as frequent but just as prominent as deed restrictions, may also be contained within the title to the property you are interested in purchasing. An easement is an agreement between the developer or property owner and another party or parties, for the use of a portion of the property for specific purposes. All easements must be honored by the present, as well as past and future property owners. In most cases easements only affect air rights for electrical, telephone, and other transmission lines, and rights below the ground surface for water, sewer and underground utilities. In other cases, however, surface easements may be established for drainage or even pedestrian or vehicular traffic or access to other property. In those cases, a portion of the property you are interested in may, and most likely will be used as a thoroughfare or subject to maintenance personnel and equipment.

ZONING—Of all the written restrictions that may be placed on a parcel of land, zoning is the most critical, not necessarily because it is more restrictive than deed restrictions or covenants, but because it is supervised and enforced. Unlike deed restrictions, however, zoning requirements being by-products of the political system are subject to periodic changes or updating. Since zoning tends to protect or increase land values, changes in single-family residential areas or the restrictions that govern them are not common. When zoning districts are proposed to be changed, sufficient notice must be given to the surrounding property owners. Such notice is usually in the form of legal advertisements in the local newspapers. Most jurisdictions also post the property with signs, advising the general public and stating the nature of any requested changes. In all instances, a public hearing is necessary to give affected persons the opportunity to voice their concerns, before a final decision will be made.

Zoning varies throughout a jurisdiction by planned use of property, by neighborhood or by section. A single-family residential zone restricts the land use in that particular section to one dwelling unit, for use by one family, on each subdivided lot. Usually the single-family residential zone is the most restrictive of all zoning classifications within the jurisdiction, for specific development standards. Consequently, property values and costs are initially higher than other residential zones and are more likely to continue to increase each year.

Single-family residential zones will not permit the construction or use of buildings for two-family or multifamily (apartments, townhouses, condominiums or patio homes) purposes. In addition, most zoning ordinances will prohibit the further subdivision or resubdivision of single-family residential lots.

Duplex or two-family residential zones allow for either a one or two-family dwelling on each subdivided lot. Multifamily residential zones in turn allow for single-family, two-family and multifamily buildings to be erected and occupied. With the exception of resort areas and cities with a very high population density, the property values and costs for lots in multifamily zones are usually much lower than single or two-family zones. In addition, many jurisdictions allow limited commercial uses to conduct business from within multifamily zoning districts in contrast to single and two-family zones, which if at all, may allow home occupations with very rigid restrictions. Zoning is a definite asset. Unless you are guaranteed neighborhood and land use stability through other means, it would be foolish to build on property that is not zoned.

In most cities and counties, zoning falls within the range of responsibilities and duties of a zoning administrator, a staff planner or the building inspection department for the jurisdiction. In any case, the person you contact is there to assist you. Do not ever feel that your questions are too minor or that you are imposing on the individual's time. Be prepared, precise and direct. Ask questions that apply to the specific lot or parcel you are interested in and also about the zoning classifications of surrounding properties.

Check for any pending or potential zoning classification changes in and around the prospective neighborhood. Also make sure you understand the use and growth potential of the neighborhood and any other area in the vicinity that may affect your investment and property values.

In the zoning research process, you will need to pay particular attention to requirements that will affect the future as well as the present development of the property you are interested in. Such requirements include but are not limited to the following items.

Setbacks or the minimum distance your structure must be located from all property lines. Setbacks vary in dimension from zone to zone within the local jurisdiction. The application of setbacks also varies, based on the nature and intent of the zoning regulations for the specific zoning district. In all cases, the dimension for the setbacks start at the front, rear or side property line and is measured inward toward the center of the lot. The setback line that is created will be parallel to each lot line, and connect to at least one other setback line. After all setback lines have been established, the buildable area, which is the maximum area within the property that may be covered with one or more structures, becomes defined and visible. The buildable area is also known and referred to in some areas as the building pad.

Fencing regulations (usually applying to height and opacity). For the most part, fence height and opacity (the amount of visibility through the fence) usually apply to the front yards only. Such regulations, if in existence, allow for total privacy for the side and rear yards, while maintaining visibility from the front by limiting height or opacity. Zoning ordinances differ by jurisdiction from no fence restrictions at all in the front yard to an established maximum height. In most cases, the fence height may be increased if the opacity is decreased (height/opacity ratio). For instance, a fence having opacity of 100 percent may be limited to three and one-half feet in height, with a provision that for each ten percent of openness, or decrease in opacity provided in the fence, an additional six inches may be added in height. Therefore, a fence with only ten percent of opacity and 90 percent openness would be allowed to be eight feet in height. The height/opacity ratio allows for security fences and large decorative front entrances without restricting your neighbors' front yard visibility in either direction.

Tree protection and landscaping. Tree protection is rapidly becoming a major concern in many communities and subdivisions across the country. The concern is so great in fact, that zoning regulations are being enacted almost on a daily basis, to prevent indiscriminate "clear cutting," or the wholesale removal of healthy trees with any size to them. Under regulations of this sort, only trees that are diseased, dead or within the areas of the house, driveway, pool, etc., will be allowed to be removed. Under-brushing and thinning of trees is also permissible but within the limits of the ordinance. Trunk diameter, height and shape and the proximity to other trees are usually among the criteria that must be established before permission will be granted for additional tree removal.

Parking and vehicular access. Minimum parking requirements for residential zoning districts usually specify that enough area be reserved on site to park between one and three vehicles. Some zoning ordinances require all vehicles including cars, noncommercial trucks, campers, vans, motorcycles, motor homes and boats to be inside a garage or screened from visibility. Some zoning ordinances require campers, motor homes and boats to be screened from visibility but allow cars, noncommercial trucks and vans to be parked in full view, while other zoning ordinances allow all vehicles to be parked in full view. All zoning ordinances require access to residential property be from a public or private street. To qualify as a recognized access, however, a private street must have a recorded permanent easement.

GENERAL—On completion of investigation of the title and zoning regulations and only after you are convinced that you can build within their guidelines, consider the remaining factors.

Check for the availability of water, sewer, electric, telephone, cable TV and gas. If any of those services are not already installed and accessible to the site, you will most likely have to pay construction costs and experience possible time delays before they will be provided.

Scrutinize the visibility from the lot in all directions. Your home site should have a pleasant view in all directions, therefore, you need to spend time on the property viewing all surrounding land for as far as you can see. Bear in mind that dense summer foliage can screen undesirable property or objects that may become visual nuisances from the fall to the spring of each year when the foliage is gone. View all surrounding properties, existing uses and buildings from all areas of the proposed building lot. Pay particular attention to existing or proposed roads and highways, large buildings, or zoning districts adjacent to the neighborhood, which would allow industrial or moderate to heavy commercial development. In addition, you should also consider noise, odor and air pollution from such uses as airports, highways, factories, or processing plants. Such uses could cause the problems listed even at substantial distances from the residential site.

Observe the neighborhood tenant mix to determine the balance or ratio of residential uses to nonresidential uses. To maintain property value for your personal single-family residence, avoid building in an area where more than 50 percent of the residential units are rented.

Check for the amount and payment schedule of fees or assessments for sidewalks, curb and gutter, street paving and driveway connections, if they do not already exist.

Check the proximity of the property to schools, churches, parks, shopping, hospitals, etc.

Determine if mail delivery service is available in the neighborhood.

Determine the jurisdiction that will provide police and fire protection and garbage pickup.

Check the condition of roads, direction and flow of traffic, amount of traffic at various times of the day, dead end streets and cul-de-sacs, signs and other traffic controls.

Determine grade elevations. Request a topographical drawing or data documenting the grade elevations of the property. That information is necessary to determine drainage and areas requiring fill or excavation. If the property is adjacent to a body of water, which is subject to flooding, grade elevations are required to establish the height above grade for the first floor under the *Federal Flood Control Act*. In addition, the proposed building site can be compared to street and adjoining property elevations.

Count the number and note the size and species of existing trees. Trees add protection from high winds by acting as nature's shock absorber. In addition, they provide shade in the summer and, after the leaves fall, allow passive solar warmth in the winter. Finally, the natural beauty of trees creates the perfect setting for any architectural style.

Locate the water table or level of subsurface water. The water table should be checked in all cases when a basement is included in your plans. A high water table will get even higher during heavy or constant rainfall. The water table could range from inches to several feet below the ground's surface. A high water table may cause water damage or other moisture related problems to building components as time goes by.

Check the lot size. Be sure you select a lot with a size that will suit your needs. An excessive amount of land may cause high yard maintenance expense and consume a large amount of servicing time. A lot with a small buildable area may not allow for the house size or layout you desire. Remember that a single story house design requires twice the land area for construction as a two story house with equal square footage (floor area).

LOT PURCHASE—Prior to purchasing a parcel of land for your proposed residence, the seller must be capable of delivering a clear title (a deed to the property which is free of mortgages, judgments or liens of any type) to you as the purchaser. To determine if the property is free of encumbrances, a title search will have to be performed by an attorney. The cost of the title search is the responsibility of the seller. A clear title should be available before you commit yourself to the purchase. Two methods of tying up or holding the property until the title search is completed are:

Offering a contract for the purchase of the property contingent upon the delivery of a clear title; or,

Depositing the agreed upon purchase price for the property in an escrow account, to be paid to the seller upon delivery of a clear title. If you select the escrow option, elect your bank or attorney to be the third party or agent that physically holds the money. Regardless of the method you choose, however, do not under any circumstances put any of your money into the property for improvement purposes until the sale is complete.

Before you can obtain a construction loan or permanent financing for your residence, the lending association will require you to own the property and be able to deliver a clear title or have a long-term lease that does not expire before the mortgage is paid. In many cases, if the time period between the date of closing for the property and the date you apply for your financing is relatively short (within several days), the lending association may accept the title search prepared for the previous owner. If you know your approximate construction costs or have a maximum budget for your project at the time you purchase the property, you may be able to arrange a purchase closing and loan closing on the same day, thus saving the cost of a title search.

Since you do not have to pay interest on a construction loan until you make your first "draw" (the point at which you start using the money), you will have a little cushion of time to begin the project. To keep a good rapport with your lender, however, try not to extend that cushion any longer than necessary.

Other than professional real estate developers and investors, few people believe that residential property can be purchased at a bargain price. Substantial savings can be realized by taking your time and using a few tried and true techniques.

KNOW THE PRODUCT—Research the real estate market in the neighborhoods that you are interested in. Research can be accomplished by reviewing the most recent property transactions recorded in the office of the Registrar of Deeds or Clerk of Court for the county in which the neighborhoods are located. In addition, many local newspapers print property transactions on a weekly or monthly basis. By obtaining information on recent transactions, you can establish reasonable market values and draw comparisons to the property you are interested in purchasing. From the information you obtain, you can determine whether the asking price for the property is high, low or average.

While performing your market research in the office of the Registrar of Deeds or Clerk of Court, you can also search for the title of the present property owner. By locating the title issued to the present owner, you can determine several important factors that will assist in your negotiations.

First, and most important, you should be able to determine the actual price paid for the property by the existing owner. If the existing owner is the original subdivider or developer, or if the property was obtained as a gift, by will or estate, a selling price will not be given. Armed with that information, you will know the amount of profit the existing owner is attempting to make and the range in which you can negotiate.

Other bits of information that you can determine, and may be useful include, the length of time the existing owner has had the property, who the prior owner was and reference to easements, deed restrictions, covenants, original plats or subdivision maps and specific property or mineral rights for which other parties may have a claim. Property title records are public information and the assistance of the Registrar of Deeds and/or the Clerk of Court is free in most areas. Document copies may also be obtained for a nominal fee.

PROPERTY APPRAISAL—Wise real estate shoppers today request the seller or agent to produce a certified real estate appraisal for the property being offered for sale. A certified real estate appraisal is a property value report from a disinterested professional third party land appraiser. Very often real estate brokers will request a real estate appraisal before listing the property for sale. Once completed, the appraisal reflects the fair market value of the property.

As a prospective purchaser, you have the right to review any appraisals that may have been performed for the parcel of land you are considering. If the property is priced in the range of the fair market value, the seller will be more than happy to allow you to see the report. If an appraisal has not been performed, you are well within your rights to request one if you are seriously considering the purchase of the property. Such a request may also be made as a contingency to a purchase proposal or contract.

By establishing the fair market value for the property, you can determine whether the asking price is high, low or average. If the asking price is higher than the fair market value, the appraisal report will give you a valid reason and increase your opportunity to negotiate a lower price. The result of that single negotiation may very well amount to several thousand dollars in cash savings.

If the property that you are interested in happens to be in a resort area or other choice location, the seller may balk at obtaining an appraisal. Usually in such instances, the seller knows the local market and realizes that if you do not purchase the property, someone else will. In that particular situation, you will have to judge for yourself whether your wants and needs for the unevaluated property outweigh the asking price. Property, which is in high demand, usually sells quickly at the asking price without the seller having to make any concessions, including a real estate appraisal.

In all cases, the expenses incurred in obtaining a certified real estate appraisal are the responsibility of the seller. If the property owner refuses to produce an appraisal or if you prefer to assess the property before you start negotiations, a quick and fairly accurate appraisal can be obtained at the local tax assessor's office. Obtain the legal description for the property and consult the city or county tax records. Not only can you determine the value placed on the property for tax purposes, (which usually has a value ratio based on a millage rate, which can be calculated into real value) but, based on the selling price, you can also determine what your taxes will be. The tax records are public information and the assistance of the tax assessor is free in most areas.

LENGTH OF TIME ON THE MARKET—The longer a parcel of land remains on the market, the better your chances are of negotiating a lower price. In most instances, a parcel of land remains on the market for a long period of time because it is priced higher than the fair market value. If the seller earnestly wants to move the property and the longer it remains on the market, the person will become anxious or in real estate terms "motivated." The more motivated a seller becomes the lower the price you may be able to negotiate.

A seller may become motivated and have to sell the property for a variety of reasons. One reason naturally is the need for quick cash.

Other reasons are for business ventures, moving away from the area or to settle an estate. Whatever the reason though, chances are very good that, with a little bit of serious negotiation you will be in position to make an excellent purchase.

In some cases, property remains on the market for reasons other than high price. In every instance, before making an offer to purchase, make sure you determine the reason the property has not sold. The most common reasons other than high price, for property to remain on the market for a long period of time are:

Poor economy;

Development saturation, or overbuilding of residential uses in the area (if you really like the property, this factor can work to your benefit);

Very poor or no access to the property;

No public utilities or service to the property;

Off the beaten path or in a remote area;

Swampy, poor drainage, or wet land (If you have suspicions that the parcel has poor drainage or is wet land, inquire or check the property yourself during or after a heavy rainfall. The best times of the year to check for poor drainage and wet land are in the spring and fall.);

Unusual size, shape or topography;

Proximity to undesirable land uses (airports, factories, etc.);

Is in an area of declining land values;

No clear title.

With the possible exception of the first two items, a fantastic deal is still no bargain if one or more of those conditions exist.

TAX SALES—One of the best ways of obtaining a parcel of land at a very reasonable price is by purchasing it from a government entity (city, county, township, state) at a delinquent tax sale. Most of the undeveloped property offered at tax sales is well off the beaten path and very expensive to develop or in blighted and rundown areas. Every now and then, however, a very good and sometimes even choice building site is offered. Surprisingly, not very many people attend tax sales, and even fewer submit bids. Those people that do submit bids quite often walk away from the sale with parcels of real estate for a fraction of their actual value.

If you can afford to invest the time to investigate the delinquent tax rolls for the area that interests you, it may pay off in substantial savings on the purchase of a lot. Tax rolls can be found at the municipal, township or county tax collectors office. You will find the tax collector and his or her staff more than willing to assist you and answer any questions that you may have. You just may find a building site that suits your needs perfectly.

MULTIPLE LOT SHOPPING—Once you establish your wants and needs for a building site, shopping for it should become as comparative as shopping for any other consumer product. When you shop for an automobile, television or even a pair of shoes, you compare different brands for features, style, color and a variety of other things before you purchase. Just as you would not purchase the first automobile or television that you looked at, likewise, you should not purchase the first parcel of property that you see. Look around and know what is available in the price range that you can afford and the general area in which you are planning to build. As a matter of fact, it is not foolish to negotiate for several parcels of land at the same time, nor is it foolish to let the sellers know that you are doing so. If the seller knows that you are seriously shopping and fully intend to purchase a parcel of land, that person will be more inclined to negotiate a lower price with you.

SALE BY OWNER—Reading the classified advertisements published in your local newspaper or even a drive through the residential areas of your city or county will reveal a number of lots or parcels of land that are for sale by their owners. Negotiating directly with an owner has it distinct advantages, such as:

Not having to pay a real estate sales commission above the final negotiated price of the property;

Quicker and more direct response to an offer from you to the seller;

A broader range to negotiate within; and,

Firsthand knowledge of the history and usability of the property.

In major metropolitan areas and more densely populated cities and counties, booklets and newspapers are published exclusively for persons attempting to sell property on their own. Those publications can usually be found in local retail establishments and quite often in racks in shopping centers, supermarkets or restaurants.

When dealing with an owner/seller, the procedure for determining fair market property value established above will assist you in the negotiating process. By establishing the fair market value of the property before you negotiate the price, you can set a maximum amount that you would be willing to pay and silently inform the owner that you are aware of local land costs and purchasing techniques.

MAKE AN OFFER—Regardless of whether you are dealing with a real estate agency, an owner/seller or even a subdivision developer, look at the property, gather as much information and as many facts as possible and then make an offer. Do not, however, make a verbal offer on site. Go home, review the information you have gathered, decide if you would like to purchase the property and then make a written offer in contract form. Do not be afraid to make an offer substantially below the asking price and even below the fair market value. The very worst thing that can happen is that you will receive a flat refusal. More likely than not, you will receive a counter proposal by the seller. At this point the negotiations have begun.

Remember that no contract is complete until the seller of the property and the purchaser agree to the terms and sign the document, although offers and counter proposals can go on for some time without either party making a commitment. Making an offer in contract form with a cash down payment as "earnest money" informs the seller that you are serious and ready to purchase. In all cases, cold, hard cash is very difficult for a seller to turn down.

An offer to purchase is not a complicated document and does not require either an attorney or a real estate broker to prepare. You can prepare an offer to purchase yourself and save the professional fees that may be involved if prepared by a third party. In all cases, an offer to purchase must contain specific information. If you prepare the document yourself, make sure you include the following items:

The full name of each person who will be identified as the purchaser on the property deed;

The full name of the seller (present owner) of the property;

The "legal description" (lot number, block number and subdivision or section) of the property, including the town, city, township, county and state (if the property is in acreage and not distinguished by lot, block and section, use the existing deed description or surveyors identification);

The offered purchasing price;

How payment is to be made;

Condition of the property (This item may be used to require the seller to leave the property in the same condition it is in on the day the instrument is signed or to make certain alterations or improvements as a contingency of the sale. Bear in mind that you are purchasing the property as a building site and will be laying out money for preparation before construction. The more preparation and/or cleanup that is required, the more money you will ultimately pay. It is critical, therefore, that the property be delivered to you in as clean and usable condition as possible. All dead or dying trees and shrubs should be removed or replaced if they are outside the buildable area and all trash, garbage or debris of any kind should be eliminated by the seller.);

The requirement for the seller to produce a clear title;

An expiration date for the offer to be accepted;

Other conditions of sale, if any;

Only contract (this item will convert the offer to purchase to a sales contract upon signing by the seller and purchaser);

Possession date of property (Possession is normally at the time of closing. In some cases the seller may desire the use of the property for a period of time after closing. In such a case, you should charge weekly or monthly rent.);

Insurance (only if the property has a good number of mature or decorative trees and shrubs that may be damaged in the event of fire or natural disaster);

Conveyance;

The date of the document preparation; and,

A place for the signatures of the purchaser, seller and at least two witnesses for each.

Upon completion of the preparation of the offer to purchase, you should either hand-carry the document (preferred if at all possible) or send it by certified mail (return receipt requested) to the seller. In all instances, before you offer a proposal to purchase, make sure you are satisfied that you have investigated the property thoroughly and that you are getting the best possible deal. Do not overlook such items as reviewing the tax obligations (not only for the property but also for the approximate value of the house you intend to build on it) for the city and county in which the property is located and the local soil conditions. Be especially skeptical if the property has been filled. Determine the area and depth of the fill and the material used. Any materials such as trees or wood, trash or any other substance that will rot and gradually compact in time can be the cause of major structural damage to your house as settlement occurs. If you have even a remote suspicion that the land has been filled, insist on a "soil test" prepared by a competent soils engineer before consummating a purchase.

The following document is a completed example of a simple "Offer to Purchase and Sales Contract" containing all of the criteria listed above. It may be reproduced and used as is or it may be used as a base for a more complex document if one is needed. The words that are shown in Italics correspond with the information that you will provide.

This document constitutes an Offer to Purchase and Sales Contract for the specific parcel of land as described in paragraph one. It is the only offer under consideration and is being made subject to the terms and conditions stipulated herein. The stipulated terms and conditions, however, may be amended upon mutual agreement of all persons stated by name in paragraph one. To constitute mutual agreement, an amendment to this document must be in writing and must be signed by each person stated by name in paragraph one, with at least one additional person, who is not directly involved with the transaction, to affix his or her signature as a witness.

1.   We, *John T. and Mary R. Jones*, hereinafter referred to as "Purchaser," offer and agree to purchase, and *Tom G. and Alice J. Smith*, hereinafter referred to as "Seller," agree to sell that certain parcel of property known as Lot *10*, Block *10*, of the *Shady Valley* Subdivision, located at *100 Walnut Grove Lane*, in the City of *Franklin Heights*, *Kingston* Township, *Jackson* County, State of *South Carolina*.

2.   The full purchase price agreed upon by the Purchaser and the Seller shall be $*50,000.00*, to be paid to the Seller by the Purchaser in cash, certified check or cashiers check at the time of closing.

3.   It is further agreed that the Seller shall remove prior to closing, all trash, garbage, debris and dead, diseased and dying trees and shrubs. The Seller shall not remove or allow to be removed, any earth, healthy trees, healthy shrubs or otherwise alter or modify the property as it exists at the time of signing of this document.

4.   This offer to purchase is being made contingent upon the Seller producing a clear and marketable title to the property described in paragraph one of this document, free from encumbrances and encroachments of any kind.

5.   This offer shall remain valid until *January 10, 2008*, at which time if not accepted and signed by the Seller, shall become null and void.

6. It is further agreed that when signed by the Purchaser and Seller and duly witnessed, this document shall become a contract to purchase the property described in paragraph one and, that this contract expresses the entire agreement between the Purchaser and the Seller and shall be enforceable by either party by specific performance and that there is no other agreement, oral or otherwise, modifying the terms and conditions contained within.

7. The Seller shall deliver to the Purchaser at the time of closing, a clear and marketable title along with any referenced plats or documents in a form, which is adequate for recording or registration.

8. The Seller shall also supply, pay for and have affixed the appropriate deed excise tax stamps to the title.

9. The Seller shall pay any sales tax that may be required for the sale.

10. If necessary, to enable the Seller to make the conveyance as herein provided, the Seller may, at the time of closing, use the purchase money in whole or any portion thereof to clear the title of any or all encumbrances or interests, provided that all instruments so procured are promptly recorded, and the intent to use the purchase money for this purpose is fully disclosed to the Purchaser before this Offer is accepted.

11. Possession of the property described in paragraph one, by the Purchaser, shall be at the time of closing.

12. Until the time of closing, the Seller shall maintain fire and hazard insurance in the amount of 25 percent (*$12,500.00*) of the purchase price.

13. The Purchaser shall be allowed full access to the property for the purpose of inspection of its condition any time between sunrise and sunset on any day prior to the time of closing.

14. This document was prepared, signed by the Purchaser and delivered to the Seller on *November 31, 2007.*

15. This offer to purchase agreed upon by the Seller, and signed into contract on *December 1, 2007.*

16. Signed and Sealed in the Presence of:

Paragraph 16 is actually the signature section of the Offer to Purchase and Sales Contract. It should contain as many lines or spaces as necessary to allow for the signatures of all of the persons stated as owners on the existing title, of all of the persons who will be stated on the new title as owners after the sale is consummated and at least two witnesses. Typewritten names must also be provided below the line or space for each signature.

If you happen to be negotiating for several parcels of land, do not (unless you can afford to purchase more than one lot) make an offer on more than one parcel at a time. Your offer to purchase is a contract and you are agreeing to buy the property. As stated earlier, do not be afraid to make a substantially lower offer than the asking price and negotiate to your advantage. The chances are excellent that you will purchase the property at or near the price you want to pay.

You will realize the greatest amount of savings on any single parcel purchase by negotiating directly with the property owner. Negotiating with subdivision sales representatives is a whole lot tougher, mainly because the subdivider or developer fixes the lot prices. The best time to purchase property within a subdivision is as soon as possible after the developer puts the parcels on the market. You will find that at that stage the developer is interested in getting a number of the first parcels sold and developed. Early sales and construction activity within a new subdivision gives the impression to other prospective purchasers that the project is successful and, therefore, a desirable place to invest. Many subdividers will offer additional amenities or "incentives" to early purchasers to "jump start" the sales. The more activity, the greater the sales potential for the remaining parcels. If you are fortunate enough to find property that you like within a new subdivision, negotiate early. By doing so, you may save several thousand dollars by being an initial purchaser and may also gain additional amenities.

When selecting a building site for your proposed house, a little shopping can save you a lot of cash. Shop wisely, but by all means bear in mind that in the selection and purchase of property, cheapest may not always be the least expensive if the parcel has major development or natural problems. Plan for function and privacy of your proposed residence by first defining the site needs. By properly defining site needs and listing priorities in advance of purchasing your property, hidden costs as well as personal aggravation can be kept to a minimum, if not eliminated altogether.

Before purchasing a lot, a Site Selection Checklist should be developed to provide a visible list of features and elements to be considered. Unfortunately the size and formatting of this book prohibits the inclusion of useable checklists. The following text, however, contains all the elements necessary to develop a comprehensive Site Selection Checklist. For simplicity and ease of use, you may prefer to eliminate the elements that do not apply to the property you are considering. The words that are shown in Italics correspond with the information that you may include on your checklist.

A Site Selection Checklist can be as comprehensive or as sparse, as you desire. In all instances the *lot number, block number, subdivision* or *section,* the *address, city, township, county* and *state* are necessary elements. The *lot dimensions* and *zoning classification, required setbacks for all yards* and the *type* and *number* of *easements* that are on the property are also important to list.

In many subdivisions *deed restrictions* apply and *Architectural Review Boards* may function. If so they should be included on the checklist. *Maximum allowable building height, minimum floor area,* the *total number of trees,* the *number of diseased, dead or dying trees,* the *number of trees to be removed* and the *number of trees to be protected,* if required by deed restriction or zoning should be included. Also consider *landscaping* that may be required. If fence regulations are contained in your deed restrictions or zoning ordinance, you will have to include *maximum fence height* and *opacity* for all yards.

Most zoning ordinances require *off street parking requirements,* and *screening* for *automobiles, vans, noncommercial* and *commercial trucks, motorcycles, motor homes, campers* and *boats.* If the zoning ordinance governing your property requires parking and screening, they should be included on your checklist. You should also include, *access to property* provided by *public street,* or *private street.*

The overall neighborhood is a major concern for maintaining property values. In that respect, the amount or percentage of *commercial* to *residential occupant mix,* and the amount or percentage of *neighborhood property rental* should be included. *Minimum distance required between structures* on the lot and any *proposed zoning changes* should be considerations.

The availability of utilities serving the property is important to know in advance, therefore, your checklist should include the status of public service for *water, sewer, electric, telephone, cable television* and *gas.* Visibility and view are critical for resale. Unless you have compelling reasons to exclude them, you should include the *visibility* and *view* from the *front, rear* and both *sides* on the checklist. Include the *zoning classification* for all *adjoining property.* To be aware of any potential costs after you purchase the property, the checklist should include any known or proposed *assessments* for; *street paving, sidewalks, curb and gutter* and *driveway connections.*

Neighborhood convenience elements such as *distance to schools,* your *church,* the *park, shopping* (convenience and mall) and *hospital* should be included. Also include *mail delivery, police protection, fire protection* and *garbage pickup.*

Other general elements to include on the list should be *street condition, lot drainage, amount of fill* or *excavation required, previous fill installed* and *the type of material used,* if a *soil test* was ever performed and the *results,* the *average grade elevation,* if the property is in a *flood zone* and if so, the *flood zone classification* and *required base flood elevation* above mean sea level. You will also need to include whether the property is *wet, soggy* or *holding water* and the *adjoining property elevations* on the *front, rear* and both *sides,* if a *plot plan* is available and the depth of the *water table.*

To complete your checklist, include information such as *clear title, appraisal performed,* the *appraised value,* how the property is offered for sale, such as by a *realty company, owner, tax sale, estate sale, other.* Also include the estimated annual *property tax* for the *city* and/or *county,* the *asking price* for property and the *fair market value.*

*If approximate value has been established.

CHAPTER TWO

PLAN SELECTION

House designs have evolved through time to serve a variety of needs and desires. Originally, the greatest need was for shelter from the weather. In addition, a sound well-built house provided a physical barrier for protection from wild animals and intruders. Once the necessary requirements of protection were satisfied, the desire for greater comfort and privacy developed. With each generation, more elaborate amenities were included until today's necessities were yesterday's luxuries. That trend and the eventual evolution in home design are continuing. A vast variety of styles, sizes, and layouts round out the basics for individual house planning.

Plan selection will include incorporating the elements of protection, comfort, privacy and whatever degree of luxury you desire. To accomplish that, however, will require a bit of thought and planning on your part. This chapter will address house planning and plan selection and provide you with the necessary information to plan wisely. In some instances, house plans are selected before a building site is chosen. Although this is the exception rather than the rule, it is perfectly acceptable providing a parcel of land can be found to accommodate the size and shape of the structure without modification. In the majority of the cases, house plans are selected after the building site is obtained. This chapter can be applied to either situation, but more so in the case of the latter.

SITE PLANNING—A major consideration in planning your new home should be the existing neighborhood conditions. Make every effort to keep your proposed home within the middle to upper middle price and size range of the existing homes within the neighborhood. "Overbuilding" a house in price or size may result in an initial devaluation of the structure, and a much slower appreciation rate. Keeping a house within the middle to upper middle price and/or size range, will increase the potential for resale, as well as afford a steady value appreciation. If a larger or more expensive home is necessary to suit your needs, consider purchasing a lot within a more exclusive subdivision.

OBTAINING PLANS—The two most common methods of obtaining house plans are through a designer or a plan service. In most cases, cost will determine the selection. A licensed architect is by far the most qualified person to design your home. An architect's services also happen to be the most expensive, but you benefit by getting a set of plans with a design and layout tailored to your specific needs and tastes. You also have the advantage of the architect's professional knowledge and experience in selecting the proper decorative and structural materials. Furthermore, an architect has the talent to solve site-related problems and is better qualified to produce a set of plans that will meet all local zoning and building codes. Direct job supervision, consultation during construction, preparation of contracts, bidding and obtaining building permits are services most architects offer in addition to custom designing. Architects also purchase and maintain "errors and omissions" insurance to cover any lawsuits that may occur because of structural failure to the building during or after construction.

In some communities, the services of an architect may not be available. You may also discover that in some instances the architect may have a large workload and be too busy to take on another project. If this is your situation, or if you feel that you want to save the difference in fees, a less expensive custom set of plans can be obtained through a nonlicensed, semiprofessional designer or drafting service.

Although nonlicensed, semiprofessional designers fall short on architectural precision and service, a quality set of plans can nonetheless be produced by some of them. In fact, many plan services of this type employ persons who are gaining the experience and knowledge in preparation for their architectural license examination. In other instances, you may find that the person offering the service is a draftsperson, employed full-time by an architectural firm and moonlighting to make extra money. Be extremely cautious however in your selection of a nonlicensed, semiprofessional designer. You are well within your rights and it is in your best interest to request references and to view plans and finished homes constructed from plans prepared for other clients.

The least expensive method of obtaining house plans is through a plan service. In utilizing this method, you choose a home plan from a wide selection of stock designs. The prints are then reproduced from original or master drawings and mailed to you after receipt of the plan fees. The advantage of this service is a substantial cost saving if the house is constructed according to the plans without any modifications.

Another advantage of a plan service is that most of the plan service companies are owned by or employ licensed architects. Through volume sales, the initial cost and fees for producing a set of house plans and specifications are spread out over a number of clients. This gives you the benefit of having an architecturally prepared set of working plans for a fraction of the price.

One of the disadvantages, however, is the lack of custom design. A popular plan may very well be sold and the house constructed any number of times within a city or subdivision. Another drawback is that more often than not, you will want the plan changed or modified after it is received. This is normally accomplished by employing the services of a designer. If this is the case, the money saved initially is used to produce the changes. Modifications to a stock set of plans may also invalidate any warranties the plan service company may offer.

Another point to consider is that design related problems that may occur during construction (yes they do happen) are much more difficult to address with a plan service company. At very best, you will be able to discuss your problem by telephone with the actual designer or person in charge of the plan service company. Under normal circumstances, however, you will speak with a customer service representative that will not be able to address the problem. That will lead to more telephone calls and probably one or more mailings to and from the plan service company. The worse case scenario is that you will need to hire a local design professional to provide a method of correction. Needless to say, delays due to design problems during the construction process could be very expensive.

Finally, the plans produced are done so under the sanction of a building code that may not be recognized in the area in which the house is to be constructed. This situation could cost more money in the long run and result in a loss of precious time if it causes a delay in the issuance of the building permit until plan corrections could be made.

PLAN NEEDS—The house plan you select should not only reflect your taste but also your life-style. Emphasis should be placed on the floor plans to provide the best traffic flow and circulation possible. Your needs and desires in style and interior layout may differ greatly from that of your friends and relatives. During the design process you will no doubt receive many suggestions as to how your house should look or be laid out. You will be the person paying for the construction and occupying the house, so by all means allow your own interests to prevail.

Poor planning in house size is one of the most common mistakes made today. House size should meet your immediate needs with strong considerations given to the future in three, five-year increments. This will allow for a total planned time span of 15 years. Bear in mind, however, that homes can be oversized as well as undersized. If you plan for family growth over the allotted 15 year time period, utilize a house plan with an expandable layout. Remember that storage area should be increased in direct proportion to other floor space. An expandable layout must also be initially situated on the property to allow for the future additional land coverage. Expansion must be planned for at the time the house is originally designed to avoid possible zoning or subdivision problems when the plan is continued.

If you anticipate a decrease in family size within the 15 year period, plan a house that will serve your needs without expansion. You can also select a home design with sections that may be closed when not in use. Such a plan would be practical for the small family that may entertain more than the average number of overnight guests (Parents or grandparents with large families or persons living in or near a resort area). Select your floor plan to suit your everyday needs.

Most design professionals plan space for sleeping, living and recreation, with buffers between each. Once the desired interior space is provided for and the exterior perimeter defined, the architectural style can then be considered or applied. The least expensive house that can be built is the basic rectangular shape with gable roof. The addition of sections or "wings" to the house and high pitched, unusual shaped roofs can run the construction cost up considerably.

An economical method of increasing interior area without affecting the roof or exterior configuration, however, is by use of a two-story design. A two-story house design will allow a greater amount of interior space, while occupying a smaller area of your building site. A full two-story house design will utilize half of the roof and foundation areas than a single story house design of equal floor space. In construction dollars, the cost of the roof (including the framing, sheathing and covering) and the foundation systems for a full two-story house will provide you an approximate 40 percent savings for the two systems.

Careful planning of the bathroom layouts in the design stages will also result in a saving to you during construction. Whenever practical, bathrooms should be placed back to back or one above the other. Such configurations allow the plumber to use less material (water supply lines, drain, waste and vent piping and fittings) and installation time.

HOUSE LOCATION—The very first step in developing a set of house plans is the preparation of the plot plan. The plot plan is a drawing usually prepared by or from information obtained from a land surveyor. A properly prepared plot plan should be drawn to scale and contain several important items of information, including but not limited to the following:

Dimensions of the property;

Meets and bounds of the property;

Dimensions of all structures presently existing and/or proposed for the site;

Required deed and/or zoning setbacks;

Adjoining streets, bodies of water or other natural boundaries;

Contours or grade elevations (the height of the property above mean sea level at several points which illustrate high, low or sloping areas and where they occur on site);

A legal description or other identifying information concerning the location of the site; and,

A north point or north arrow.

The location of the building on the property is very critical if your home design incorporates either an active or passive solar heating system or water heater. Be sure to specify if any type of solar system is to be used before the plot plan is prepared. Location of the solar collection components on site in relation to the sun, existing or planned landscaping and the area required for the equipment will have a definite and distinct effect on the operation or efficiency of your system.

PLAN CONTENTS—To assure proper construction within the shortest amount of time and with the least amount of confusion, a complete set of detailed plans is necessary. Naturally the definition and contents of a complete and detailed set of plans will vary greatly from designer to designer. Nevertheless, since you are footing the bill, you should insist on at least the following sheets.

Site Plan: The site plan is a scaled sketch of the property showing how the house will be situated on the lot. The site plan should also show the relation of the streets, setbacks, landscaping, swimming pool, driveway, garage and other accessory structures to the house.

Note! In many instances a properly prepared plot plan may be substituted for a site plan.

Floor Plans: Floor plans consist of one or more sheets showing all the heated and unheated floor area in and around the house. The basement, first floor, second floor, etc., garage, carport, patios, decks and porches are typically shown on the floor plans. Floor plans contain such information as room layout and dimensions, window and door locations, equipment location and exterior building dimensions. A variety of additional information will also be found including location of all walls, openings, arches, half walls, closets, stairs, plumbing fixtures, mechanical equipment, electrical devices and specific finishes or materials.

Elevations: Elevations show the interior and exterior walls of the building, as they will appear from all sides or views. Elevations may tend to be confusing if the house has projections or a varied floor plan. Since elevations as drawn for construction purposes are only two dimensional, the interior walls and exterior perimeter of the house are shown on a flat, nonprojecting basis. It will be much easier to understand the elevations (especially exterior) if you view them in conjunction with the floor plans. Other information that should be shown on your elevations include the finish materials, dimensions for door and window heights, floor elevations, chimney heights, exterior grade elevations, roof breaks and pitches, porch elevations in relation to interior floor and exterior grade elevations and in some cases landscaping, if your house incorporates the use of planters or provides for trees and shrubbery in the overall design.

Foundation Plan: In most cases, the foundation plan will look rather skimpy compared to the detailed floor plans and elevations. The information contained on that sheet, however, is extremely important. In no case should the foundation be installed without strict adherence to the plan. Bear in mind that the foundation is the only component of the house that cannot be replaced! The foundation plan will contain such information as the length, width and depth of the excavation (known as the footing); the size, depth and location of the fireplace and pier or column footings; the size, spacing and location of reinforcing steel if needed; unusual soil conditions, drainage specifications (including foundation or "French" drains); and, the exterior wall line of the building. Smaller (toe) footings will also be shown for porches or nonloadbearing walls.

Wall Sectionals and Details: A wall sectional is a view through a wall or section of the house from the foundation, to and including the roof. This cut away view will show sizes and types of materials, how they are applied and their direct relation to other materials. Fastening devices, anchors and other material ties will also be shown or designated, as well as interior and exterior finishes. Sectionals also show the depth below grade of a basement if one is to be installed. Perimeter drainage, filled and compacted earth, stair sections, and other specific and important information will also be included on the sectionals.

Additional details may be included on any of the sheets with the possible exception of the plot plan. Additional details may include window and door schedules, materials and finishes for elements on the interior elevations, plumbing, electrical and mechanical layouts with risers, and trim details.

Electrical Plan: An electrical plan shows a wiring diagram for the entire house. It shows the location of all switches, receptacles, light fixtures and other devices or components that will be part of the electrical system. An electrical plan will also specify the calculated electrical load for the house, the wire types and sizes, a schedule of electrical fixtures, a schedule of electrical symbols used and details about the electrical service equipment and connection. Comprehensive electrical plans also include wiring for low voltage wiring such as telephone, cable television, alarm, intercom, computer and related systems.

Plumbing Plan: A plumbing plan shows all of the piping for the hot and cold water supply and drain, waste and venting systems the entire house. It shows the location of all fixtures and other devices or components that will be part of the plumbing system. A plumbing plan will also specify the piping types, sizes and materials, a schedule of plumbing fixtures, a schedule of plumbing symbols used and details about the water and sewer service and connections

Mechanical Plan: A mechanical plan shows the location of the supply and return ductwork for the entire house. It shows the location of all mechanical equipment, dampers (if used) thermostats and other devices or components that will be part of the mechanical system. A mechanical plan will also specify the calculated heat loss/gain for the house, the sizes and location of all supply and return registers, a schedule of mechanical equipment and components and a schedule of mechanical symbols.

Specifications: Specifications may also be included with your plans, either on the drawings themselves or as a separate document. Specifications contain information concerning the project in general, such as responsibility for obtaining permits, insurance requirements, responsibility for obtaining subcontractors, etc. The specifications may also contain a variety of information for construction of the building such as special materials and components that will be used, equipment brand names and models, the building codes used for design purposes and any special requirements due to location, soil conditions, etc. Specifications may be very comprehensive or very broad in content and in some cases not even included with the plans.

Stick to your plans! The time to modify any area of the house is in the planning stages. Modifications during construction waste valuable time and are very costly. Costlier yet are changes made after basic components are in place. For instance, a simple change to enlarge a room after the framework is installed will require the framing to be removed, then reinstalled in its new position. You pay to have the wall built in its original position, you pay again to have it removed and you pay a third time to have the wall reinstalled in its modified position. As you can see, even minor changes can result in a terrible waste of money. Several changes on a single job can result in severe budget problems, not to mention the waste of time and materials.

Your home will be your domain. Design it as such. Plan for areas and rooms within your house to be more than bare minimum. Try to attain maximum livability without going overboard on design. Keep circulation and movement direct and simple. To save time and money, develop or select your plan based on your needs and desires. Develop a checklist to determine those needs and desires and use it as a guide to selecting the plans for your dream house. The following text contains all the elements necessary to develop a comprehensive Plan Selection Checklist. For simplicity and ease of use, you may prefer to eliminate the elements that do not apply to your specific property or needs. The words that are shown in Italics correspond with the information that you may include on your checklist.

Your Plan Selection Checklist should include information such as the *name, state registration number, mailing and E-mail addresses, telephone* and *fax* numbers and *fee* for your *architect* or *the name, mailing* and *E-mail addresses, telephone* and *fax* numbers and *fee* for your *drafting service.* Critical elements for the proposed house that should be on the list include the *height* and *area* of the initial house, the *expandable height* and *area* (if part of the design). Also include the *total estimated budget* (if known at the time) for the house and the average *value of existing homes* in surrounding area.

Since the primary concern of the Plan Selection Checklist is the house design and its associated features you will certainly need to include (as applicable) the *interior layout, exterior design, traffic flow, expansion area allowance, closable areas, energy efficiency, solar* or *special equipment, view* and *landscaping.*

The complexity of your design can range from the simple to the sublime. The sheets required for your residence and your checklist, therefore, may include a *plot plan, site plan, floor plan(s), elevations, wall sectionals, details, foundation plan, basement plan, drainage plan, framing plan, electrical plan, plumbing plan, mechanical plan* and *specifications.*

CHAPTER THREE

EXPECTED COSTS

Now that the site and plan selections have been completed, you are ready to start the third and probably the most critical phase of construction planning. That phase, being the establishment of your expected costs and budget, will determine the size, complexity and types of finishes for the house that you can feasibly build. In this chapter I will not only point out the costs that must be budgeted, but also the areas where reductions can be made without sacrificing the integrity, structural stability or appearance of your home.

COST ANALYSIS—Your cost analysis may be prepared for construction costs only or for the entire project. If prepared for construction only, your analysis will be limited to the costs associated with the erection of the building. An entire site analysis will anticipate all costs associated with the project, including amounts paid for the land, plan fees, insurance, etc. as well as the actual construction. The example used in this chapter will be based on the entire site analysis.

The key to saving money during construction is to develop a very rigid budget management program and adhere to it without exception. Self-discipline and control must be practiced to the hilt! Once the decision to use a specific plan, certain type of material, component or subcontractor is made, stick to it. Cost overruns are generated in almost every case as a result of changes made to the original plan.

PROPERTY COSTS—The purchase price of your building site will be the very first cost you will be subjected to. In calculating your true purchase price, do not forget to include all additional fees you are charged in excess of the actual land cost. Such additional items, such as deed recording fees, attorney's fees, etc., although small compared to the actual amount paid for the property, still result in money spent and must be considered part of the overall land cost.

Survey: At some point in time a survey of your property will become necessary. It would be most beneficial to you financially if the survey can be performed before you purchase the property. If you purchase within a new subdivision, chances are the original corners (usually iron pipes or solid rods) will still be visible. If the corners are not visible, a survey will be necessary to establish your property lines. Since you, as the prospective purchaser, have the right to know exactly what you are negotiating for, the expense of a survey, which is usually very nominal, should be the responsibility of the seller. Insist on the survey at the time you offer a purchase agreement, as a condition of the sale. If the property owner is serious about selling the parcel, a survey fee will not stand in the way.

Plan Fees: Although the fee you pay for obtaining a set of building plans is not a direct construction cost, it should nonetheless be included in your overall construction expenses. Plan fees could be included in your construction budget either before or after they are purchased. If you budget for your plans in advance of receiving them, however, do not be tempted to exceed the amount of the original figure unless absolutely necessary.

Clearing: If the lot you decide to purchase happens to be heavily wooded, some clearing may be necessary before you can start construction. Bear in mind that healthy trees add to the initial, as well as resale value of a residential lot. In the clearing process, only plan to remove trees that are diseased, damaged, poorly shaped or are within the area of development.

Good Advice! You should also consider removing all softwood trees such as pine or fir, that may be close enough to cause damage to your proposed house if blown down or broken due to the additional loads caused by snow, ice or wind. If your lot contains large hardwood or softwood trees in close proximity to the onsite "footprint" of your proposed house, be aware that during the excavation for the foundation, major roots may be severed. Once major roots are severed, the affected trees will require removal. If not removed you run the risk of one of two situations occurring. First, the trees affected will become weakened and top heavy due to reduced support to the root system. In that situation, the trees become much more vulnerable to uprooting under the load of a moderate to heavy wind. Second, severed roots cut off the source of nutrients to the remainder of the tree. In that situation, the deprivation of nutrients eventually leads to the death of the tree. Needless to say, a dead tree with any size to it can cause substantial damage if it falls on a house (yours or your neighbors). A simple method of determining if a tree is in jeopardy of suffering major root damage is to look at the "spread" of the tree's foliage. The spread is the maximum width of the leaf system. The spread of the root system below the ground is approximately the same as the spread of the leaf system above. If you have concerns about the location of major root systems or potential damage to existing trees, consult with a tree expert. If you need to hire a consultant, don't forget to include the fee in your cost estimate.

Excessive tree removal or "clear-cutting" (removal of all trees on the site) will reduce the value of the completed project. Do not clear-cut unless your site plan calls for substantial decorative landscaping, replenishment of the removed trees or because it is absolutely necessary for development purposes. Clearing may be performed in a variety of ways with varying costs. Listed below are the most common methods of clearing.

Hire a contractor who specializes in tree removal. A tree removal contractor will provide the equipment and manpower to remove the trees and stumps. This method is normally the quickest and most expensive. The trees, however, are removed by experts who know how to get the job done in the shortest amount of time and with a minimum amount of damage to surrounding vegetation or property. In addition, a tree removal contractor will either remove the trees and limbs entirely from the property or remove the limbs and saw the trees into fireplace length logs. A professional tree removal contractor will also be covered by liability insurance for protection in the event of damage to your neighbors' property. When negotiating with a tree removal contractor, insist on proof of liability insurance coverage.

Hire nonprofessional tree cutters. Although this method would cost less, more often than not, the persons would not have the equipment to do a quality job. Nonprofessionals also have a tendency to leave the tree stumps in place. They too will either remove the trees and limbs entirely from the property or remove the limbs and saw the trees into fireplace length logs. Nonprofessional tree cutters usually do not have liability insurance coverage. Lack of insurance may not be a problem for you, however, unless you have a small lot, plan to remove trees close to your property line or have large trees that may fall across your property lines when cut.

Cut the trees yourself. This method may not save you much money unless you are experienced in tree cutting. If you do not already own the equipment for cutting and dropping the trees in the proper place, you would have to rent or purchase it. In addition, removing the stumps will require other specialized equipment, knowledge or back breaking work. You can elect to cut the trees yourself and hire a subcontractor to remove the stumps, but the difference in cost per tree may not be enough to justify your time or the danger involved. Unless you are very experienced in tree cutting, I do not recommend this method.

Give the trees to others. Often you can find people in your general area that will cut the trees for you at no charge, in exchange for the wood. People who offer this service usually will not remove either the limbs or the stumps. Like nonprofessional tree cutters, the people who offer free tree removal quite often do not have liability insurance coverage.

Sell the trees. In most areas, lumber and/or pulp mills will buy healthy trees. With this method, experienced tree cutters employed by the mill remove the trees. A good quality job can be expected and some but not all of the limbs will also be removed. The balance of the debris and stumps would have to be removed by others, but the money you will be paid for the trees will defray if not totally eliminate the cleanup costs. Naturally, this is the most recommended method for tree removal. The sale of your trees, however, is dependent upon the number of useable trees you have to sell and demand for raw wood by the mills at the time you are ready to clear your lot.

Filling: After the property is cleared to your specifications, raising the ground elevation or filling of the lot may be required before construction can start. That change of elevation is normally accomplished by adding clean fill dirt to the areas to be raised. The most common method of obtaining fill dirt is naturally to purchase it. In some cases, however, clean fill material can be obtained free of charge. Check all the neighborhoods within a reasonable driving distance of your building site. Look for other building sites where excavation is taking place. The dirt that is being removed could very easily be hauled to and deposited on your site. Other possibilities would be excavation for swimming pools, utility projects, etc. A telephone call to the Public Works Department of your city or county and to specialty contractors or subcontractors (excavation, pool, etc.) may very well result in you getting all your fill material for little or nothing.

Insurance: Although insurance rates are relatively small in comparison to other costs you incur during construction, don't let the fact that they are nominal or intangible lessen the importance of a good policy. Construction insurance or what is commonly known as a builder's risk policy is necessary to keep you out of the courts. Your insurance should cover you and any other person who may be on the job site by invitation or otherwise. The insurance policy that you purchase should not be in lieu of policies required to protect your subcontractors or any of their authorized personnel. Do not permit any subcontractor to talk you into purchasing its required insurance policy. As a matter of fact, proof of insurance should be made a requirement of your written contract (further information on this subject will be covered in Chapter Six). For your own protection, your policy should include personal injury, fire on the structure and material and theft on the materials and tools.

Permits: In almost every instance, permits for construction of your home will be required. Depending on your specific area requirements, permission for construction of the entire job may be granted in a single building permit. In most cases, however, separate permits will be required for building, plumbing, electrical, mechanical (heating and air conditioning), gas and possibly others. Each permit issued carries a fee with it. Check with your city or county government in the early planning stages to determine the permit fee schedules, if any. The fees you pay for your permits are required to process the paperwork and defray the cost of inspecting your job site. In some cases, building inspection departments charge for reinspections if initial inspections fail. Unless they can be justified by building code or city or county ordinances, no other permit fees should be paid.

Excavation: Excavation costs normally involve the removal of the soil in preparation for the footings and a basement if one is to be installed. Removal of soil can only be accomplished by two means; mechanically or by hand. In some cases, however, both methods may be necessary. Removal by mechanical means is by far the fastest and easiest method. It also happens to be the most expensive, since heavy equipment and qualified operators will be required. Unless optimum conditions exist, the excavation by heavy equipment would still need "dressing" or cleaning up by hand. Excavation dressing may be performed by a laborer or even by you, since the equipment operator will remove virtually all the excavated material.

The remaining alternative would be to remove the soil by hand. This is usually the least expensive, but most time-consuming method. A point to consider, however, is if your footing is deep or basement area vast or if the soil happens to be unusually hard or contains large amounts of damp clay or rocky material, you would more likely save money by utilizing the heavy equipment. Analyze the soil conditions carefully and get several bids for equipment and hand excavating before proceeding with soil removal.

In most instances, the soil that is removed by the excavation process can be used on other sections of your property (fill, landscaping, berms, etc.). If you are absolutely sure the excavated soil is more than you will need to complete the project, sell it or use it to barter for other materials or labor that will be necessary at a later time. Do not give it away! Do not allow a subcontractor or heavy equipment operator to haul it away unless you receive credit for it based on present market value per load. Finally, do not, under any circumstances pay to have it removed. If push comes to shove, put an advertisement in the classified section of your local newspaper and sell it.

Construction: The actual construction phase of cost estimating will consume the largest amount of your budget. Several acceptable methods exist which will produce accurate cost figures. Choose the method that suits your particular needs from the options below. You may want to use two or more of the methods as a means of crosschecking the accuracy of your initial cost calculations.

Architect: If an architect is commissioned to produce your plans, for an additional fee he or she will also conduct the bidding process for the job. To conduct the bidding process, the architect will solicit bids from contractors qualified to construct the house according to the plans and specifications. In most cases, the architect will select the contractor submitting the lowest bid. You as owner must then approve or disapprove the selection. The major benefit in using this method is that the architect assumes more of the initial duties and responsibilities. This process, however, will cost more, take you out of direct initial contact with the contractor and may even take you out of the roll of the contractor. Remember that the purpose of this book is to save money in the construction processes. The largest percentage of savings is realized if you assume and maintain the role and responsibilities of the contractor.

Bids: You may also conduct the bid procedure for the purpose of determining costs. Obtain as many bids as you think will be necessary, but at least three, from reputable contractors. From all of the bids received, eliminate the highest and lowest, and any others that do not show a reasonable range in consistency. Average the remaining bids and eliminate 20 percent.

Example: You may receive bids from six contractors that range as follows; Contractor "A" $80,000, Contractor "B" $105,000, Contractor "C" $106,000, Contractor "D" $106,500, Contractor "E" $120,000 and Contractor "F" $122,000. Eliminate contractors A, E and F (high and low bids) and add the remaining bids of contractors B, C and D together. Divide the total (317,500) by three (for the number of bids being averaged) for a figure of 105,833. Multiply 105,833 by .20 (the percentage of savings you can realize only if strict adherence to the principles of this book is maintained) for a figure of 21,166. Subtract 21,166 from 105,833. The result, $84,667, will be an accurate construction cost estimate.

Note: The dollar amounts used in the above example are for the purpose of illustrating the process and do not represent actual or current construction costs. Due to a variety of reasons, construction costs differ greatly on a nationwide basis.

TABLE 1

Bids received from six contractors;   Eliminate and average;
Contractor "A" $80,000   Eliminate
Contractor "B" $105,000   Add
Contractor "C" $106,000   Add
Contractor "D" $106,500   Add
Contractor "E" $120,000   Eliminate
Contractor "F" $122,000   Eliminate

Eliminate contractors A, E and F and add contractors B, C and D

Add $105,000
                                            106,000
                                           +106,500
Total $317,500

Divide total by three   3)317,500

                                            105,833
Multiply answer by **.20   x .20
                                             21,166

                                            105,833
Subtract from average   -21,166
                                             84,667

$84,667 will be an accurate construction cost figure.

**20 percent only if strict adherence to the principles of this book is
maintained.

Professional Estimator: Professional building cost estimators can now be found in most metropolitan areas in the country. A professional estimator is a person who, for a fee, will review a set of building plans and specifications and render a very accurate cost estimate for any phase of construction, from a single system to the completed project. In the case of an estimate for a project owner, a complete estimate is preferred. Professional estimators stay informed and up to date on material and labor costs in the general areas in which they work, on a continuing basis and can produce very accurate estimates. Be aware that in all but very unusual instances, professional estimators are either employees of or derive the bulk of their business from, homebuilders or general contractors and may have a backlog of projects that stretch from several weeks to several months.

Since the estimator bases the project cost exclusively on your building plans and specifications, the more detailed information you can provide, the more precise the estimate will be. Remember that the better estimators are employed by or kept busy by contractors and will give their employers or most frequent clients the highest priority. If you have the time to wait for one of the better estimators, the estimate you receive will be detailed, professionally prepared, extremely accurate and well worth the fee charged. You may also be fortunate enough to locate an estimator that is employed by a contractor, who would be willing to work on your plans as a personal (after hours) project. Most contractors that employ estimators, however, will not allow them to perform estimating services after hours or on a private basis.

Some estimators only estimate material costs and do not include labor costs. Make sure you understand the full extent of the service that will be provided before you hire a professional estimator. In addition, determine if the estimator is strictly independent and not affiliated with any material supplier. Professional estimators that are affiliated with, or have a strong preference for any particular material supplier, may get a supplemental fee for business referral. As with all individuals or entities you hire, check references. To locate professional estimators in your area, check the yellow pages under the main heading "Construction" and the subheading "Estimating" or "Estimators."

Computerized Estimating Program: If you are proficient with a computer and have time to spare, you may want to look at several of the consumer oriented construction estimating, home design or Computer Assisted Drawing (CAD) programs that are on the market. I must stress that if you select this option, look at consumer oriented programs only. Unless you are planning to actually become a professional designer or estimator, stay away from commercial CAD or estimating programs. Commercial CAD and estimating programs are very large, very complex and difficult to master. Once the frustration of attempting to learn the program finally sets in, its presence on your computer will do nothing more than gobble-up a massive amount of hard drive space. In addition, commercial CAD and estimating programs are extremely expensive (much more than the fee for a professional estimator) and require you to learn far more than you will need to perform an estimate on a single project.

The basic function of a computerized estimating program is to calculate the amount of material necessary to build the structure. The computer will apply a unit price per component and produce material costs for individual systems (framing, electrical, plumbing, etc.). Once the system costs are determined, they will be added together to obtain the total material cost for the entire structure.

Now for the obvious question, "where do the material prices come from?" The answer is, you enter them. That means you must call various material suppliers and get the current prices for every component in every system to get an estimate that is useable. Bear in mind that material costs fluctuate on a regular basis so one hundred percent accuracy is not guaranteed and entering material cost information for a single project could be cumbersome. Unlike the commercial estimating programs, most consumer-oriented programs will calculate the material cost only. Some advanced (and more expensive) consumer programs will also calculate labor costs, but you must obtain and enter those figures as well as the material costs before the final calculation can be performed.

If you believe you can learn and apply a computer program effectively and save money in the process, then this option may suit your needs. If so, select your program wisely and remember that the estimate will only be as valid as the accuracy of the information you enter into the computer.

Material Suppliers Estimates: In most cases, material suppliers will give you extremely accurate figures on quantities and cost of components. This method may be the ideal for the vast majority of residential projects. When requesting this service from potential suppliers, refer to it as a material "take-off." A material take-off may come from a variety of sources. Listed below are the most common materials and the most likely sources to check.

Concrete: Pre-mixed concrete plants.

Lumber: Lumberyards, building supply stores.

Electrical components: Electrical supply stores, some building supply stores.

Plumbing components: Plumbing supply stores, some building supply stores.

Mechanical components: Mechanical equipment supply stores, some plumbing supply stores, some electrical supply stores, some building supply stores.

Note! The mechanical subcontractor normally supplies mechanical equipment. In most areas, however, the local Power Company will assist in the sizing of the heat and air conditioning units by calculating the heat gain and loss for your house. This service is usually free of charge, but don't take it for granted, check first. Don't be surprised if your mechanical subcontractor recalculates the system. It certainly does not hurt and may turn up discrepancies in the first set of calculations.

Hardware: Building supply stores, hardware stores.

Brick and Masonry Units: Brick plants, some building supply stores.

Other Materials: Building supply stores, hardware stores.

Material suppliers normally estimate quantities of materials based on accepted industry formulas. The following is a list of the most common materials and the formulas for calculating quantities. You may use the formulas to do an initial material take-off or as a crosscheck of estimates submitted by material suppliers.

Concrete: Calculate the total square feet of the area to be covered; multiply the total square feet by the thickness of the component; (footing, floor slab, etc.) convert the answer to cubic feet and divide the cubic feet by 27 to determine the total cubic yards.

Standard Brick: Calculate the total square feet of the area to be covered; multiply the total square feet by 7 (brick per square foot) to determine the total number of brick.

Oversized Brick: Calculate the total square feet of the area to be covered; multiply the total square feet by 5.3 (brick per square foot) to determine the total number of brick.

Brick Pavers: Laid with the face (wide area of the brick) up; calculate the total square feet of the area to be covered; multiply the total square feet by 4.5 (pavers per square foot) to determine the total number of pavers.

Standard 16" Block: Linear feet of wall divided by 1.33 equals number of block per course.

Note! In calculating brick (or block) for walls, do not omit the areas for windows and doors. Calculate all walls as if they were solid.

Lumber: Length (in feet) multiplied by the width (in inches) multiplied by the thickness (in inches) divided by 144 equals board feet.

TABLE 2

BOARD FEET

Lumber Dimensions	Board Feet per Lineal Foot
1 inch X 4 inches X 12 feet	.333
1 inch X 6 inches X 12 feet	.500
1 inch X 8 inches X 12 feet	.666
1 inch X 10 inches X 12 feet	.833
1 inch X 12 inches X 12 feet	1.000
2 inch X 4 inches X 12 feet	.666
2 inch X 6 inches X 12 feet	1.000
2 inch X 8 inches X 12 feet	1.333
2 inch X 10 inches X 12 feet	1.666
2 inch X 12 inches X 12 feet	2.000

Wall Studs: One per linear foot of interior and exterior wall based on 16" centers.

Plywood: Calculate the total square feet of the area to be covered; multiply the total square feet by 32 (4' x 8' sheet) to determine the total number of sheets required.

Note! This formula may also be used for any other material that is supplied in 4' x 8' sheets.

Additional 2" x 4" Framing and Bracing Lumber: Total linear feet of interior and exterior walls times eight.

Note! This calculation will provide the additional lumber needed in linear feet. The actual individual lengths you purchase will depend on where the lumber will be used. Calculate each area separately and try to purchase lengths as close as possible to the dimensions needed.

15 Pound Felt (roofing underlayment): 432 square feet per roll.

30 Pound Felt (roofing underlayment or rolled roofing): 216 square feet per roll.

Roofing Shingles: Calculate the total square feet of the area to be covered; divide the total square feet by 100; (square feet of coverage pet square of shingles) add 1 square foot for every linear foot of valley and ridge to determine the total number of squares of shingles.

Dry Wall: Calculate the total square feet of the floor area; multiply the total floor area by 3.5; divide the answer by 48 (4' x 12' sheet) to determine the total number of sheets required. An alternate calculation is; calculate the total square feet of the area of the exterior walls and the area of the ceiling to be covered; calculate the total square feet of the area of the interior walls to be covered and multiply that answer by 2; add the total area calculations for all walls and ceiling areas to determine the total area to be covered; multiply the total square feet by 48 (4' x 12' sheet) to determine the total number of sheets required.

Note! In calculating dry wall material for interior and exterior walls, do not omit the areas for windows and doors.

Floor Covering: Calculate the total square feet of the area to be covered; divide the total square feet by 9; to determine the total square yards.

Wall Covering: Calculate the total square feet of the area to be covered; divide the total square feet by 9; to determine the total square yards.

Paint/Stain: Depends on the application. Check the coverage rate on the can for the brand you prefer to use. Bear in mind that coverage rates differ for primer, sealer and finish coats.

Cost Index: This method is the easiest, but also the least accurate. It gives a good "ball park" figure, which can be used for quick general estimating. Refinement of the figures, however, will be necessary for construction purposes.

To use this method, you must first establish the average cost per square foot for residential construction in your area. Next, you simply multiply that figure by the total square footage in your proposed house and you have your total cost.

All cost index calculations are based on average values for average homes. If your home has an unusual shape, a complicated roof design or more than the average interior amenities, then additional monetary considerations would have to be made. The figures derived by this process will show the material costs plus labor and profits. Bear in mind, however, that the estimated figure will be reduced further by as much as 20 percent.

If upgraded materials, components or fixtures (usually finish materials and appliances) are substituted after construction has started, the savings ratio will fall based on the amount of the total over expenditures. In such cases, the cost savings for the materials and the labor for installing them will remain the same. The higher prices paid for better quality or additional materials, components or fixtures, however, will add expense to those individual items that cannot be considered part of the initial overall savings ratio.

Example: To obtain a construction cost estimate using a cost index figure, multiply the total area of the house in square feet by the cost index figure. The result will provide a total valuation in dollars for the finished house. Multiply the valuation by the percentage of the savings factor you anticipate (20% is used for this example). The result will provide the total construction cost savings in dollars. To obtain your actual construction cost figure, subtract the cost savings from the total valuation.

TABLE 3

Cost comparison for three houses in the low ($65), medium ($80) and high ($100) price ranges:

House Size 2,000 square feet of heated space.

Total Area in Sq. Feet	2,000	2,000	2,000
Cost Index Figure	x $65	x $80	x $100
Valuation in Dollars	$130,000	$160,000	$200,000
Savings Factor	x .20	x .20	x .20
Total Savings	$26,000	$32,000	$40,000
Original Valuation	$130,000	$160,000	$200,000
Less Savings	$26,000	$32,000	$40,000
Your Cost	$104,000	$128,000	$160,000

Note: The dollar amounts used in the above comparison are for the purpose of illustrating the process and do not represent actual or current construction costs. Due to a variety of reasons, construction costs differ greatly on a nationwide basis.

The cost index figure (construction cost per square foot) may be obtained in a variety of ways. The most reliable source is the building inspection department serving your particular area. Other good sources are architects, homebuilders, general contractors, building material suppliers and independent cost estimators. The reference section in your local library is also a good source if it stocks current trade oriented documents.

OTHER COSTS—Although there are unanticipated costs that usually appear during the course of any project, most if not all of them are predictable. Along with the major costs listed above, other expenditures may very well materialize. Examine the following list of other possible costs and then do whatever research you think is necessary to determine whether or not they will affect your project.

The purpose of this book may be greatly inhibited, if not defeated altogether, unless all costs are considered initially.

Approval Fees: Some subdivisions and local zoning ordinances require approval of a review board before a building permit can be issued. Such boards may review your plans for compliance with deed restrictions, zoning regulations, building size or appearance. If for some reason your house cannot be situated on the lot without a violation of your setback requirements, (the distance you must place the house from the property lines) an appeal for a variance to the local Board of Adjustments may be necessary. Be aware, however, that a variance approval is not an automatic or even easy experience. In the majority of jurisdictions that have boards with the authority to grant variances, the appeals must be presented as "hardship" cases. Before you attempt to appear for a variance, make sure that your request is legitimate. Your idea of a hardship may differ greatly from the board's definition. If for instance you happen to own a parcel of land that has an unusual shape or one that is smaller in size compared to the majority of the others in your neighborhood or one that has unusual topographical conditions or any other situation which, being the fault of the property, would reduce your buildable area to the point that you would not be able to develop the lot to minimum neighborhood standards, then a request for a variance would be in order. On the other hand, a hardship based on design of the structure, cost or value or any other controllable factor would be termed "self imposed" and not valid.

Fees for appeals are established individually by each governmental jurisdiction. Fees can range from zero to several thousand dollars. Check your local building inspection department or clerk's office for the fee required in your area.

Utility Deposits: Deposits will be necessary for all utility services required for your home. All deposits will have to be paid in advance. Establish what services will be necessary (electric, gas, water, sewer, cable TV, telephone, etc.) in the planning stages and include the deposits and connection fees in your budget. Check the local utility offices for the deposits required.

Tap Fees: If water and sewer service is provided for your lot, tap fees may be required in addition to deposits. The tap fees may range in cost from 25 to several thousand dollars each. The fee is required for the privilege of having the service provided and normally covers the cost of construction of the lines and taps and the installation of the meter. Check your local water works or public works administrative office for the tap fees required.

Impact Fees: In addition to tap fees, some communities charge what is known as impact fees. Impact fees are more common in very popular areas or areas of rapid or high growth. The amount of the fee is based on the impact the additional water use or sewage processing will have on the overall system, which is either overtaxed due to massive use, obsolete or quickly becoming so. The fees are charged to defray the costs of future improvements to the water or sewer systems.

Septic Tank And Water Well: If your lot does not have water and sewer service available, a septic tank and/or water well would have to be installed. Tank cost, installation, well drilling, etc. would add to initial cost.

In most cases the local health department will require a "percolation" test before a septic tank permit will be issued. The cost of the test will be the responsibility of the property owner. Check the health agency with jurisdiction in your area for percolation test fees and approved septic tank and water well installers. You can also check with the prior property owner to determine if a percolation test was performed in the past. Soil conditions do not rapidly change, therefore, a percolation test should be valid for many years.

Soil Test: If for some reason you or your designer have the need to question the stability of the soil on which your home is to be erected or if the building official in your area requires it, a soil test would have to be performed. Such a test would determine the density, composition and value of the soil. A soil test is highly recommended if the property was filled at any time in the past with good, well-packed material. A soil test is critical if your property has had a history of being unstable, low or swampy or if it was filled with an undesirable material. Undesirable materials would include large bulky items that will create voids upon back filling as well as items such as wood, paper, trees, junk, trash, etc., which would decompose or rot after a time period. A soil test can only be performed by a person or organization qualified to do such work and a fee will be charged. If there is no listing in your telephone book for a soil-testing agency or service, check with an architect, engineer or your local building official for companies or individuals that are approved to perform such tasks.

Curb Cuts And Street Connections: If your property has curb and gutter at the street and it will be necessary to remove a section of it to accommodate access, then a curb cut will be required. If a curb cut is required, payment of a fee may be necessary. Bear in mind that the streets in your area may be under the jurisdiction of one of several governmental agencies including city, township, county, state, and federal. Establish the agency concerned and then contact the responsible person to obtain permit or fee information.

Portable Toilet Rental: In most areas, the health agency or building inspection department require clean toilet facilities be provided on or near all construction sites. In areas where toilets are not required, they should nonetheless be provided for the workers. Single stall units are rented by the month at a nominal fee. Your budget should have an allowance for toilet rental for a minimum of six months.

Trash Removal: Your local jurisdiction or subdivision may also require a method for the removal of construction trash. If so, you will need to obtain a waste container from a local waste management company. Waste containers are available for rent in a variety of sizes. The rental fee is on a monthly basis and determined by the size container you need. In addition to a monthly fee, some waste management companies also charge for container delivery and a dumping fee. The dumping fee is based on the trash weight, volume or both.

To get a true valuation of the total project the expenditures must be broken down into two components, land cost and construction cost. The land cost includes all expenditures that would be associated with the property if it were to remain in its undeveloped state. The construction cost includes all expenditures required for the placement of your house on the property. Only the construction costs will be considered, however, when applying for a construction loan. Depending on the policy of your lender, the land may or may not be accepted as collateral for the construction loan. Your property must be fully paid for and free of encumbrances of any kind before applying for a construction loan.

Anticipation of all costs for construction is very critical. Make sure you establish a budget that accurately accounts for all possible costs. Develop a checklist for cost estimating and try to anticipate every possible outgoing dollar. The figures used for checklist purposes must be the lowest negotiable or final costs. The following text contains all the elements necessary to develop a comprehensive Cost Estimating Checklist. For simplicity and ease of use, you may prefer to eliminate the elements that do not apply to your specific property. The words that are shown in italics correspond with the information that you may include on your checklist.

Your checklist should start with the actual *land costs*, which will consist of the *land purchase price* and the *survey cost*. Since land improvements will be included in your construction costs, do not consider them as part of your *total land costs*.

Next list all of your actual construction costs. Construction costs should be broken down into five subclassifications.

The first subclassification should be the *land preparation costs.* Include items such as *clearing, tree removal permit, excavation, excavating permit, filling* and *street access.* Also include *percolation test, soil test, curb cut, street connection* and *appearance/appeal board* if required. The total of those items will give you the *total land preparation costs.*

The second subclassification should be *utilities* and include items such as w*ater well, septic tank, water tap fee, sewer tap fee, water service deposit* and *sewer deposit.* Be sure to include *electric service deposit, gas service deposit, cable TV service deposit* and *telephone service deposit.* You may be required to pay a *water impact fee* and/or a *sewer impact fee.* Include several spaces for *other* and a space for *total utilities costs.*

The third subclassification should be *construction permits* and include such items as *building permit, plumbing permit, electrical permit, mechanical permit* and *gas permit.* Include several spaces for *other permits,* a space for *plan review* and a space for *total permit costs.*

The fourth subclassification should be *other costs* and include such items as *plan preparation fee, insurance, toilet rental, trash removal service,* several spaces designated as *other* and a space for *total other costs.*

The last subclassification under construction costs should be *materials and labor costs* and include the items *materials* and *labor.* Also include a space for *total materials and labor costs.*

To arrive at a total estimated project cost, create a column with the heading *total estimated costs* and list all of the costs you stated on the checklist so they can be added. The column should include, in successive order, the *total land preparation costs, total utilities costs, total permit costs, total other costs* and *total materials and labor costs.* The result will be the *total construction costs.* Add the total construction costs to the *total land costs* and the result will be your *total estimated project cost.*

CHAPTER FOUR

SHAMS, SCAMS, CONS, SWINDLES

AND FRAUD

As discouraging as it may sound to a potential owner/contractor, fraud, deceit and an abundance of con-artists and swindlers thrive in the construction industry. That is not to say legitimate and quality oriented tradesmen are hard to find. They are in fact plentiful in most areas. It only means common sense and reasonable judgement must be applied before hiring your subcontractors.

Note! In the vast majority of cases, subcontractors are good, honest, hardworking people. When reference is made to subcontractor(s) in this chapter, it is not intended to be synonymous with con-artist(s) or intended to be derogatory in any way. It is used as a general term to describe any person or entity offering to perform specific construction work.

Protecting yourself against the con-artists is not difficult if you look for the obvious signs during your subcontractor selection process. Some of the signs are easy to spot, others may take a little time and research on your part to determine. It is very important to remember that nothing can drain your construction account faster than a con-artist. Therefore, it is critical for you to take whatever time that may be necessary to be satisfied of the integrity and reliability of the person or company you are negotiating with, before your decision to hire is made.

The following signs will assist you in spotting con-artists or the potential for fraud. Bear in mind that this is not an all-inclusive list of potential schemes or fraudulent practices. In addition, certain signs may exist for quality subcontractors while con-artists may not exhibit any signs at all. When selecting subcontractors, your principal rule-of-thumb should be to use care, caution and common sense.

BUSINESS LOCATION—The location of a subcontractor's business can be a sign or a deception. The vast majority of subcontractors conduct their businesses from a physical location. The largest or most prosperous will usually function from an actual office location in a commercial, business or industrial zone within a community. In some instances, quality subcontractors conduct their businesses from an office located in or adjacent to their homes. Beware, however, of the subcontractor that operates solely from a motel room or post office box.

The motel room/post office box business location is usually a sign that the subcontractor is not permanent and may only be there for the purpose of taking your money without delivering a product or service. As obvious as it may seem, many people, including builders, fall for the scam often enough to make it a viable and lucrative source of income for the con-artist. The usual mode of operation is simple but extremely effective. The subcontractor arrives in town and sets up a base of operations by utilizing the street address of a local motel or an independent post office box. The street address may reference a "suite" number that corresponds with a room or unit number without mentioning the motel name. For example, if you did not know that the address is for a motel, "Joe Scammer Construction Company, Suite 307, 123 Main Street" may look like a physical location for a legitimate business. He then advertises in the local newspaper, by flyer or by personal visits to job sites. Like con-artists in any other scam, the subcontractors are convincing as well as believable with their stories of experience, quality and "this must be your lucky day" price quotes.

Although quality subcontractors do, on occasion, list a post office box as their only business address, they are very few and far between. Lack of a true physical location may be an indication that the subcontractor does not have sufficient assets or the financial ability to support a healthy business. It may also indicate that the subcontractor wants the ability to remain mobile, just in case the urgent need to leave town arises.

Good Advice! Unless the subcontractor is very well known within the area or highly recommended to you by at least two or more past and reliable customers, do not consider him for your project if he does not have a permanent address.

LOCAL REFERENCES—Lack of local references, like lack of a physical address, is a definite sign to consider in the subcontractor selection process. Fraudulent activities by subcontractors flourish in the area of owner/contractor building, because more often than not, the victim fails to check and verify local references. As far as references are concerned, I can't stress the word local enough.

The subcontractor that may be out to separate you from your money will normally provide you with a list of out of town, rather than local references. Such references, if you can find them at all, are usually nothing more than accomplices (friends, relatives or employees) of the con-artist, waiting to give you glowing endorsements of his abilities and the quality of his work. Be careful! You may be stepping into a trap.

Local references are usually much more accurate, easier to verify and contain a reasonable degree of reliability. You also have the opportunity to view the product from the exterior of the house and, in many instances, (if the homeowner is agreeable) the interior as well. If the homeowner was involved during any stage of the construction process, valuable information about the reliability and business habits of the subcontractor can also be obtained.

Good Advice! Ask for and check local references on every subcontractor you hire.

BUSINESS ACCOUNTS—To exist in the current business environment, every subcontractor has to have and maintain at least one local business account. Before you employ a subcontractor, you not only have the right, but in my opinion, the obligation to review as many of his business accounts as necessary to satisfy yourself that he is legitimate and reliable. To do so, you must request that the subcontractor provide you with a list of business oriented credit references, such as material suppliers, equipment rental stores and other subcontractors that may have been or may presently be employed by him.

When inquiring about a subcontractors account, you should request information concerning payment history, delinquencies or outstanding debts, late or nonpayment of bills, any repossession of equipment or materials or similar items of a business oriented nature. You should not request information concerning the amount of money he owes, the amount or type of collateral or security used to obtain the credit or information concerning any nonbusiness accounts. You should not be interested in or have the need to know about a subcontractor's wife's "Clothes-a-Rama" credit card or his "Happy-go-Lucky Saloon and Video Arcade" credit account. Stick to his business accounts only.

Good Advice! Check credit references but keep them limited to business accounts only.

QUICK DEALS—Be aware of laborers that travel in groups and offer to do work cheaper or quicker because they just happen to be in the neighborhood. Such groups, known as "travelers" develop bad reputations wherever they work. There is no question about it. If you hire travelers you will lose money.

The normal ploy is to offer the services of the entire crew for a lower than normal price, ask for full or partial payment either before or shortly after they start, then disappear. The work performed, if any at all, is usually very poor quality. In many instances the work has to be corrected or removed and replaced. In all instances, however, the homeowner is stuck with paying the repair bill.

Certain groups of travelers have been known to intimidate or even threaten the people they worked for. The best way to stay out of trouble with travelers is to hire only local subcontractors. More information concerning travelers will be provided later in this chapter.

Similar quick deal offers may be made by subcontractors that claim they are new to your area and willing to work for little or no profit. In such instances, the subcontractor will tell you of his need to get established in the area, build a local reference list or a variety of other believable reasons why he will work for little or no profit. Every business, new or established, relies on profit to exist and grow. A subcontractor who is willing to sacrifice profit to get a job is a very poor businessperson, very stupid or a con-artist. In either case, you do not need or want that type of person on your project. Subcontractors that tell you they will work for little or no profit, usually walk away from the job with the highest profit.

It is important to realize that every existing subcontractor started at some point in time as a new business. It is also important to realize that if some trusting people did not have faith in the abilities and integrity of those subcontractors, they would not have survived in business. In the majority of cases, before a person attempts to go into business as a new subcontractor, he or she gains the knowledge and experience by working with an established subcontractor.

Quite often, new, legitimate subcontractors will submit low bids in an effort to get established in an area. In addition, they will usually take their time and do a good quality job. Utilizing new, legitimate subcontractors can result in a good job at a small to moderate savings.

The question then becomes, how do you know the legitimate, new subcontractor from a con-artist? Check the work history of the individual(s) involved in the business with past employers and the local building official. Request and review the subcontractor's license and trade cards. Ask the subcontractor to provide you with a list of local suppliers that have extended credit and verify that accounts exist.

New legitimate subcontractors may charge less than their existing counterparts to get your work, but most often the savings will be small. Good quality tradesmen know what their competition is charging and what they can do the work for without financially jeopardizing their own businesses. In addition, new as well as existing legitimate subcontractors will obey all local and state laws affecting their trades. The tradesman that asks you to obtain required building permits, request inspections or perform any task normally considered part of the subcontractor's responsibility, as a condition of the deal, has very questionable motives. If you have any doubt or question about the quality, believability or integrity of the people you are dealing with do not hire them.

Good Advice! Stay away from quick and cheap deals, especially from unknown or out of town subcontractors.

NO CONTRACTOR'S REFERENCES—In the construction industry, general contractors provide the main source of employment for subcontractors. With the exception of handymen or tradesmen who offer their services exclusively for small projects, very few quality subcontractors work only for individuals. Be cautious of any subcontractor that does not work for or can at least produce good references from established general contractors. Quite often individuals in this category did at one time, work for general contractors but may have developed a bad name or reputation. In an effort to remain in business in their trades and within their communities, they offer their skills (or lack of them) to individuals only.

Good Advice! Steer clear of subcontractors that do not work for or cannot produce references from general contractors.

NO LICENSES—The most important instrument a subcontractor can own is a license to perform within the trade being offered. One or more license (depending on state and local laws) is essential if a tradesman expects to go into business as a subcontractor. Even con-artists realize the benefits of having licenses and strive to obtain them.

Most communities require subcontractors to be licensed as a prerequisite to obtaining permits. Some communities require a business license, (nothing more than a revenue producing instrument issued upon the payment of a fee, for the privilege of working within the jurisdiction) a local trade oriented license, (issued by the jurisdiction upon proof of qualifications, usually by examination) or a state trade oriented license (issued by the state upon proof of qualifications, usually by examination). Communities may require a subcontractor possess one or any combination of licenses before work can be legally performed.

Trade cards may also be required by a community for people actually performing the skilled work in one or more of the construction trades, in addition to the various licenses. If they are, require prospective subcontractors to produce them for your review. Good quality and well-qualified subcontractors will be proud to show off their credentials. If a person becomes offended or refuses to produce the documents when asked, remove him or her from consideration.

You can determine what licenses and trade cards a community requires by asking the local building official. Individuals that cannot produce a valid, current license and/or trade card on demand should be avoided like the plague.

Good Advice! No license, no work.

PHANTOM SUBCONTRACTORS—Legitimate as well as illegitimate subcontractors alike, actively solicit work on a regular basis. Once the word gets out (and it will) that you have a construction project in progress, you may be solicited by people claiming to represent various subcontractors.

A con used successfully in many areas is initiated when a person claiming to represent a licensed (and in many cases well known) subcontractor offers a proposal to perform specific work. The pitch is usually very polished and convincing and the person claiming to be the representative may have extensive, accurate knowledge about the subcontractor.

People employing the "phantom subcontractor" sham may have access to the subcontractor's license number as well as past and present work history. Don't be surprised. Such information is not difficult to obtain. A past or present employee, a friend of an employee or a person with only minimal research ability can very quickly obtain the license number and work history for any subcontractor. Even an individual with very little imagination and less resources can search local construction sites and determine the names of the subcontractors presently employed, where they worked in the past and future or pending projects. The most successful con-artists maintain a rather impressive history of jobs (in some cases better than the subcontractors themselves) performed and also in progress. They will even be more than willing to provide a prospective victim with a reference list of past clients for their review.

Once the victim hires the phantom subcontractor, the "representative" requests a payment then hits the road with the cash in hand. In some (rare) instances, the person may have material delivered to the job site or even start the job before requesting money. In those cases the con-artist usually receives payment for value of the goods and labor provided but has no intention of paying the bills when they become due.

To protect yourself from the phantoms, visit the subcontractors at their places of business. Meet the owners, officers or principals as well as the license holder(s) involved in the business. Don't be afraid to ask for the president, owner or person in charge. Most of the time they will be delighted to meet and talk with you, especially if it may lead to a contract.

Good Advice! Never hire a subcontractor until you meet and talk with the principal person.

CASH ONLY DEALS—The construction industry is full of small subcontractors that prefer to function on a cash only basis. Those subcontractors attempt to make "windfall" profits at every opportunity. At times the additional profits occur when the subcontractors get paid in cash, pocket the money and forget to pay their taxes. Regardless of the fact that a person on your project may intend to violate state and federal laws, providing cash payments to subcontractors may lead to problems for you as well. Paying for work by cash can cause problems with your record keeping and potentially make you liable for taxes, workman's compensation or other obligations of the subcontractor. In addition, indication that an individual may not pay taxes or keep accurate records is traits of a bad businessperson. Subcontractors that operate in that mode will pressure you to pay them in cash, especially if they are due a draw or finish their work at night, after normal banking hours or on weekends. Checks represent records and are part of a "paper trail." Maintaining good records and a solid paper trail will keep you out of trouble with the tax collectors.

Good Advice! Get written invoices for all work performed on your project, pay all bills by check and request receipts from your subcontractors.

TRAVELERS—As mentioned earlier in this chapter, a common scam, so far unique to the construction industry, is carried out by groups known as travelers (the most notorious known as the Irish travelers). Travelers are masters at deception. Their deceptive techniques are handed down from generation to generation with incentives and rewards given to the young members that learn the quickest. They live in and intentionally keep their society closed, and thus assure that their techniques and secrets are closely guarded.

The mode of operation of travelers is to visit a community, offer their services, bilk their clients and, before anyone becomes aware of the scam, leave town. They tend to work whole neighborhoods at a time, signing as many contracts as possible within a predetermined time period. Travelers may specialize in new construction, renovation work, small repairs, maintenance or any combination. Regardless of the specialty, however, the application of the scam is the same, is simple and works very nicely.

The head of the clan will approach the victim, introduce himself as a subcontractor and offer very quick services of his crew at a bargain rate. In some cases the rate is ridiculously low and in other cases only slightly less than the bids of legitimate subcontractors but enough to entice an unsuspecting owner/contractor. When questioned about why they are able to do the work at the reduced rate, travelers will give you any one of a list of believable answers, such as:

They are new to the area and need the work to keep the crew together;

They happen to be doing other projects in the area and can "work you in" (kind of a volume discount); or,

They just finished a big job and are getting ready to start another but have a few days to kill. In fact, a seasoned traveler is so smooth he can even make legitimate contractors believe his crew can perform one or more trades without having actual experience of any kind.

The group then enters into a written or verbal (usually verbal) contract with the victim to perform one or more construction tasks. Travelers, who are notorious for not obtaining business licenses or the appropriate construction permits, attempt to operate on an all cash basis to reduce paper trails and minimize, if not totally eliminate payment of their taxes. Therefore, the contract requires an advance payment, based on a percentage of the total cost of the project as good faith money, with work to start within several days of signing. Once the contract is signed (although travelers have no intentions of fulfilling their contractual obligations, which cannot be proven until after they leave town), if a victim refuses to make the initial payment before work begins, the travelers may place a mechanic's lien on the project or sue for breech of contract. Either way travelers make money from a construction project by performing little, or in most instances, no work.

They are normally based in a specific location but as the name implies, travelers move from town to town employing their unique scam. Travel is planned in advance and incorporates a route and sometimes a time period to complete the cycle. The cycle starts at the home base, moves to a specific mid point city, then back to base without visiting the same city twice. Although the family members who remain at home (women, children, etc.) claim they have no knowledge of the whereabouts of the group, the trips are too well coordinated to make those claims believable.

The groups are composed of tradesmen, usually related by blood or marriage, having varying levels of construction knowledge and/ or experience. In most cases, senior members of the clan provide the only exposure to training or experience the group members have. The education, however, specializes in the scam rather than the trades. Travelers are fairly fluent in construction lingo but minimal in actual skill.

The image projected by the travelers is a "down home" folksy type. They are not loud or flashy but rather tend to look like humble people willing to give a hard days work for modest compensation. Be very careful and aware of the fact that the grandfatherly gentleman offering his out of town crew to perform quick, cheap labor is a slick, professional con-artist lining you up for his next financial kill.

Good Advice! Do not hire out of town subcontractors that just happen to appear on your doorstep with an offer you cannot refuse.

JOINT ACCOUNT—On occasion, you may negotiate with a subcontractor that requests access to your construction or other checking accounts for the purpose of writing project related checks when you are not available. As obvious and foolish as it may seem, people, whom under any other circumstance would not even consider the suggestion, allow it to happen in construction.

Like all other schemes, the reasons given for making such a request may seem very reasonable at the time, such as:

"I may need materials at a time when you are not available to write a check;"

"I may need certain materials that are not stocked or readily available at the suppliers with whom you have accounts;" or,

"I may need to pay my workers at times other than normal pay periods."

The list could go on and on, depending on how experienced the subcontractor is at employing the scheme. In many instances the subcontractor may even try to convince you that the practice is normal by saying things like:

"All my clients do it;"

"We have always done business that way;"

"All the other subcontractors do it."

Good Advice! Never, regardless of the circumstances or reasoning offered, let a subcontractor talk you into allowing him access or otherwise authorize access to any of your bank accounts or sources of funds.

JOINT OWNERSHIP—Less common but more devastating, is a request to place your land, project or other property you own into joint ownership with the subcontractor. The scheme may be proposed by the subcontractor in lieu of an initial draw (which you should not pay anyway), as a form of security or as a gesture of good faith on your part. Do not go for it!

As usual, the proposal sounds simple and plausible. The subcontractor requests that the property be placed into joint ownership, with his name included with yours on the title. Upon completion of the work and after being paid in full, the subcontractor will relinquish ownership and any future claim to the property. The transaction is intended to appear as though you and the subcontractor will be "partners" on the project. He may try to sell you on the idea that by becoming partners, you protect each other's interests. After all, with both parties having a vested interest in the project, the owner will not default on payment and the subcontractor will definitely finish the work and do the best possible job for the most reasonable price. Right? Don't bet on it! The only interest served will be that of the subcontractor. For starters, the subcontractor is not obligated to relinquish ownership. In addition, once named as an owner on the title, the individual has specific legal rights and must be involved, in proportion to the percentage of ownership, in any decision or future transaction involving the property.

As incredible as it may sound, people do get duped into joint ownership. Those people quickly learn the true motive behind the scheme. Once the subcontractor is paid in full for his services, he offers to sell his share of the property back to the original owner. The victim is then faced with either meeting the demands of his "partner" by repurchasing his own property (usually at a much higher than market price) or sharing ownership with a con-artist. At that point the original owner may find himself in the position of having to rent or lease the subcontractor's share of the house before and during the time it is occupied. In addition, the subcontractor may attempt to force the original owner to sell his or her share to him by refusing to allow the property to be mortgaged. No transaction can occur for the house including sale, rent or lease without all of the owners listed on the title agreeing to do so.

Good Advice! Steer clear of any subcontractor who offers to be your partner or requests to be included as an owner of your project.

INSUFFICIENT EQUIPMENT—To do quality work, subcontractors must have the proper tools and equipment. In some instances, however, subcontractors will attempt to purchase or replace their tools and equipment by charging the costs to one or more projects. The costs for tools and equipment are normal business expenses and are to be paid by the subcontractor. Determine if prospective subcontractors have sufficient tools and equipment to perform all of their required tasks at the time of your initial interview. Include language in your contract to require the subcontractor to provide his own tools and equipment and do not allow him to talk you into accepting responsibility for buying or replacing them.

If your house has an unusual size, shape or design, however, special tools or equipment that are not normally purchased or used by a subcontractor may be required. If so, the cost is yours to pay. You should determine if any special tools or equipment would be necessary at the time of your initial interview with each subcontractor.

Good Advice! If the subcontractor possesses little or no tools or equipment, do not hire him.

NO CONTRACT—All to often, a subcontractor will try to convince you that a contract for the work he is to perform is not necessary. He will try to assure you that an oral agreement and a handshake is all that is required. I cannot count the times a subcontractor told me "my handshake is my contract" or "my word is my bond." The sad truth is that a handshake may have been fine for grandpa, but in the present, complex business environment, it is nothing more than a gesture of salutation that has no legal and very little moral value.

A slick talker may make you believe that in the few minutes you have known each other, a friendship or some sort of bond has occurred that makes you immune to chicanery. If you fall for that nonsense, he will then make you believe that since he is a "good old boy," your agreement to hire him is hinged on nothing more than you trusting him to perform his tasks and he trusting you for his pay.

A common excuse for not entering into a contract is that the job is too small. No job in construction is too small to have a clear understanding of the responsibilities of the parties involved. The only time you do not need a contract based on the extent of the job or its cost, is when the amount of the money involved is so insignificant, you do not care if you lose it. A subcontractor may also argue the point that he is known around town or has a good reputation and, therefore, never uses or has need for contracts. Unless you are convinced, absolutely and without a doubt, that you will not be the first casualty on the person's unblemished record, do not go for it.

No matter how detailed an oral conversation may be, of all the points agreed upon, some are usually fuzzy or even forgotten within hours. I have personally participated in meetings where oral agreements were made and, as the participants left the meeting place, could not agree as to the points actually agreed upon only minutes before. An oral agreement is subject to the interpretation and mental picture created by each person involved. In all instances, the interpretations, mental pictures and personal interests of each person are different than those of the others. A contract, however, provides a written record (that can be clarified if necessary) of the obligations of each party.

All quality subcontractors will be more than happy to enter into a written contract. If a subcontractor refuses to document what he will provide by written contract, find one who will.

Good Advice! No written contract, no work.

DEPOSIT—One of the most often applied scams in the construction industry, is to request partial or even full payment for the proposed work in advance, take the money and never show up to perform the task. The request may be called a deposit, a down payment, earnest money, good faith money or other business-oriented name. Regardless of how it is termed by the subcontractor, the most accurate word that can be used to describe agreeing to such request is stupid. Unless you have absolutely no regard for your money, do not even consider paying for work that is not performed or components that are not installed or at least in your possession.

Reasons given to you for requesting money in advance can range from "good business" to sob stories. The only person to benefit from a business standpoint will be the recipient of the payment. After the subcontractor receives money in advance, you may or may not (usually the latter) receive the goods or services for which you contracted. Sob stories, which can range from tearful pleas to ridiculous lies, are given to get money from you for sympathetic reasons. Several of the better sob stories I have heard on my own projects include attempts to obtain advance payment to:

Provide clothes or supplies for children starting school;

Provide for friends or relatives in the hospital (especially pregnant wives);

Purchase a gift for a special occasion being celebrated by a friend or relative; and,

Pay workers that have been out of work for a period of time (to keep them from walking off my job). I even had one person ask for "his profit only" so he could take his wife on vacation before starting on my project.

Just as the subcontractors are unknown to you, however, you are also unknown to them. Legitimate subcontractors agreeing to work for you will naturally want to protect their own interests. In doing so, they may ask you to verify that the funds are available to pay for the work upon completion. A request to verify funds is valid and verification can be accomplished by going, with the subcontractor, to your lender. Be sure to have a copy of the contract for the loan officer to review and only verify the availability of the funds necessary to cover the work concerned.

Requesting an advance payment is common and, if agreed to, may be devastating to your project. I cannot stress this point enough.

Good Advice! Under no circumstances, should you pay money to a subcontractor before work is started. As far as you are concerned, your construction project should be treated as a business, not a charity.

SPECIAL OFFER—A favorite and very effective scam is one that should be the most obvious. It is the special offer, deal or discount. It works well because virtually everyone wants to be singled out and given preferential treatment. It works so well and is so easy to perfect and employ, in fact, that even novice con-artists can enjoy a high rate of success. The task for the con-artist is to develop a rapport, gain your confidence and then offer you "the deal."

The scam starts when the con-artist engages you in a very simple and seemingly innocent discussion of a very general nature. During the conversation, the con-artist discovers coincidences or similarities in both of your backgrounds, which he will use to establish as the basis for the special offer. The coincidences or similarities are usually bogus, but on occasion may actually exist. For example, you may happen to mention that, you were the president of the student council in your senior year at high school. The con-artist will then give you an expression of disbelief and inform you that it so happens he was also president of the student council in his senior year at high school.

Most of the time the claim will be a bald face lie designed to create the coincident. In the instances where actual coincidences do occur, the con-artist will speak with more knowledge of the subject, giving his claim and subsequent special offer additional credibility. Once the coincidence is developed, the con-artist will propose the special offer, only because he feels a kinship to you through your common experience. The victim usually believes that the special offer is a better deal than he can get from any other subcontractor and does not bother to obtain additional bids. In some cases, depending on the beliefs of the owner/contractor and how personal the coincidence is (race, religion, national heritage etc.), the individual feels an obligation to hire the subcontractor. As soon as he is hired, the con-artist will apply one or more of the schemes (request a cash advance, all cash etc.) mentioned above.

Favorite topics used to develop coincidences are:

Hobbies, such as hunting, fishing, boating;

Mutual acquaintances or friends of friends;

A particular branch of military service or military assignments;

Membership in civic organizations;

Membership in social clubs;

Past participation in a particular sport, such as high school or collage football;

Present participation in a particular sport such as golf, tennis, squash etc.;

Political affiliations;

Religious affiliations;

Ethnic background;

Race; or,

Music.

Since most people love to be flattered, good con-artists will gladly accommodate, especially for an owner/contractor. He will tell you that he admires you for taking on the monumental task of building your own house, that you are doing something he always wished or wanted to do himself or praise you in other unrelated ways. Even though a con-artist doesn't know you, it is amazing how fast he grows to like and praise you. Believe me, the only thing your new best friend likes about you is the opportunity to separate you from your cash. Give any con-artist a few minutes time and he will learn enough information about you to justify a special offer.

Good Advice! Do not talk about personal events, situations, experiences, hobbies, background etc. to prospective subcontractors. Keep all conversation, especially before a contract is signed, on a business level. Beware of any special offer, discount or deal.

HIGH PRESSURE—Con-artists usually get you to fall for their scams by using high pressure or a sense of urgency. Their mode of operation is designed to get you to act very quickly, without thinking of the potential consequences. They know that, given a little time to think about the terms of their proposals, most people will decline them. A good con-artist will make you believe that if you do not act instantly, you will lose the golden opportunity he is offering. He will stress the urgency by trying to make you believe that he has other jobs waiting and if you do not act right then and there, on the spot, you will suffer in time, money or quality. Unless you know the subcontractor has a solid reputation and that he is in constant demand, take as much time as you need to make your selection.

Good Advice! Do not yield to pressure or urgency when selecting your subcontractors.

CONSTRUCTION MANAGERS—Construction managers have been used in commercial construction for many years. The title is given to a person who has the experience and technical ability to manage a project on behalf of a licensed individual or company. Since a construction manager works directly for the contractor or in some instances the building owner, a state license, bond and insurance is usually not required. That lack of regulation opens the door for anyone, qualified or not, to purchase a local business license and offer services as a construction manager. Consequently, a growing number of self proclaimed experts with little or no actual construction experience, but using the title construction manager, con individuals into believing that they can save them money by managing their home building projects. Beware!

Construction managers offer a fee based service to perform any or all of the tasks customarily performed by licensed contractors. The fee is usually a percentage of the total cost of the project that varies, based on the degree of involvement or number of different tasks the construction manager is commissioned to perform. As the builder of record, you will still bear full responsibility for every aspect of the project, be required to make all decisions and retain all of the liability.

If you are inclined to hire a construction manager, be sure to check the individual's credentials, experience, background and past clients very thoroughly. Most require a contract. Review it carefully, preferably with an attorney. In addition, make sure you understand how and how much you will have to pay for the service. You may find that the cost for a construction manager's service is close to the fee charged by a licensed contractor, in which case it would be wiser to hire the contractor.

Good Advice! If you consider turning your project over to a construction manager, also consider a contractor. You may find that the cost difference is not that great and you will relieve yourself of the responsibility and liability of being the builder of record.

Although con-artists thrive in the construction industry, you can greatly reduce, if not eliminate your chances of being a victim of their scams. Your best defenses are common sense, good business practices and time to think about the proposals made by all subcontractors. Use them wisely. Carefully scrutinize the people who offer their services to you. Examine their credentials, look at their past projects and check their references thoroughly.

CHAPTER FIVE

SUBCONTRACTORS

In assuming the role and duties of the general contractor, you will be responsible for hiring crews and individual persons who are qualified to perform each trade involved in the project. Since those crews and persons will be working directly for you as contractor, they will be referred to as "subcontractors."

Study each of the subcontractor roles and determine which tasks you can perform yourself or may be willing to assist in. Performance of or assistance in any job within the project will save you additional money. Many people have saved as much as 50 percent of the cost of construction by putting in their own time and applying the other principles found within this book. In every state that I know of, a property owner may perform the function of any subcontractor (including electrical and plumbing) in his own single-family residence.

If you elect to perform the tasks of a subcontractor, make absolutely sure that you thoroughly know and understand the theory and principles of the trade involved, especially if it requires inspections from a governmental entity. Do not attempt to perform any skilled task because it doesn't look too difficult or because you think you can do it. Many problems have developed in the operating (plumbing, electrical and mechanical) systems of residences because of the abundance of confidence but lack of ability of do-it-yourselfers. Worse yet, structural damage, human injury and even death has resulted (and will continue to result) from improperly installed operating systems.

For the purpose of this book, we will classify the subcontractors in two categories. The first category will be known as the "common" subcontractors. The second category will be known as the "special" subcontractors. Study the definition of each category and the responsibility of each subcontractor. Once you know the role of each subcontractor, you can establish a list of trades required to complete your project. (The checklist at the end of this chapter will assist you in that task).

COMMON SUBCONTRACTORS—Subcontractors who perform a specific task commonly associated with the project. Common subcontractors include but may not be limited to the following classifications.

Excavation Subcontractor: Prepares the site and removes the soil for the installation of the foundation (footings) and basement. The excavation subcontractor uses heavy equipment such as backhoes, front-end loaders, dump trucks, etc. In some instances, such as very unusual topographical lot conditions or heavily wooded property, excavation by heavy equipment will be impossible. In those cases, excavation by hand must be performed.

Note! The excavation subcontractor could also dig the trenches required for drainage or plumbing pipes or other utilities while the heavy equipment is on the site. This function could save you money and time as well as offer the advantage of getting all of your required excavation out of the way while the property is still in its rough state. In addition, the back filled areas would have much more time to settle before the fine grading, reducing the chance of low spots or trenches forming after the yard is completed.

Excavation subcontractors normally charge by the hour for the operator and equipment used.

Concrete Subcontractor (Also known as "Concrete finishers or foundation subcontractors"): Erects forms for footings and concrete wall systems when required, calculates the necessary amount of concrete and places (installs) it in the forms or excavated areas. Concrete subcontractors also install the vapor barrier and reinforcing wire and finish the concrete for floors (very smooth "trowel" finish) as well as driveways and sidewalks (level, but rough textured "broom" finish). Some concrete contractors will install the "batter boards" and steel "re-bar" (reinforcing steel rods).

Concrete subcontractors normally charge per cubic yard of concrete for placing and finishing.

Steel Subcontractor (also known as "Rod busters"): Installs steel re-bar in foundations, poured concrete walls or columns and grade beams if required.

Note! Not all plans or soil conditions require reinforcing steel for foundations. Investigate the stability and conditions of your soil and the load the proposed house may exert on it before electing not to use steel. You may also want to have a soil analysis performed by an engineer or testing laboratory. Every bit of your house investment will be supported by the foundation. Do not skimp on it!

Steel subcontractors normally charge per foot of steel installed or by the hour.

Drainage Subcontractor: Installs foundation drains around the perimeter of the house.

Drainage subcontractors normally charge by the job.

Piling Subcontractor: Installs piling foundation systems.

Piling subcontractors normally charge by the pile or by the job.

Masonry Subcontractor: Installs all masonry, block and/ or brick work from the footings up, to support the floor and wall systems. Masonry subcontractors will also install the exterior finish if it happens to be brick, stone or other types of masonry units, build fireplaces, arches, lay brick pavers (for floors, walks and driveways), walls, fences, etc.

Masonry subcontractors normally charge per unit for block, per thousand units for brick, by the square foot for stone or by the hour.

Framing (Carpentry) Subcontractor: Installs all framing materials, erects walls, floors (other than concrete) and roof. In addition, the framing subcontractor installs windows, doors and related hardware, exterior sheathing, boxing, interior and exterior trim, stairs and railings.

Framing subcontractors normally work for a predetermined cost figure (usually a dollar amount per square foot of floor area) or by the hour. You will save money, time and aggravation by choosing the square foot method. Be sure you know the customary rate in your area, however, before agreeing to an amount.

Electrical Subcontractor: Installs all electrical wiring, boxes, devices and related components necessary for the requirements and to the specification of your plan.

Electrical subcontractors normally charge by the device (switch, receptacle and light fixture) by the amount of wire in feet, by the size of the service (in AMPS), by the hour, any combination of the above or by the job.

Plumbing Subcontractor: Installs all water and sewer piping, plumbing fixtures and related components necessary for the requirements and to the specification of your plan.

Plumbing subcontractors normally charge by the fixture (water closet, lavatory, sink, etc.) by the length of piping in feet (usually in water and sewer work), by the hour, any combination of the above or by the job.

Mechanical Subcontractor: Installs all ductwork and mechanical equipment for heating and air conditioning using a forced air system.

Note! Electric baseboard heat units are normally installed by the electrical subcontractor and certain types of hot water or steam units are installed by plumbing subcontractors.

Mechanical subcontractors normally charge by the ton (air conditioning and split systems) or by time (total labor) plus material.

Gas Subcontractor: Installs all gas service piping, fittings and appliances and related components necessary for the requirements and to the specification of your plan.

Note! In some areas installation of gas piping and appliances is permitted to be performed by plumbing and/or mechanical subcontractors.

Gas subcontractors normally charge by the job or by time plus material.

Roofing Subcontractor: Installs felt underlayment, flashing around chimneys and in valleys, etc. and shingles on the roof. (In some cases the framing subcontractor will install the felt underlayment.) It is also common for roofing subcontractors to install rain gutters and leaders (downspouts) and certain types of siding.

Roofing subcontractors normally charge by the "square" (100 square feet of roof area covered) or by the hour for roofing and siding and by the foot (in length) or time plus material for rain drainage systems.

Dry Wall (Sheet Rock) Subcontractor: Installs and finishes the interior wall and ceiling systems (sheet rock or gypsum board).

Dry wall subcontractors normally charge by the square foot of total board area handled (not installed) or per sheet for "hanging" and finishing.

Plaster Subcontractor: Installs wet plaster over lath or masonry for interior and/or exterior finishes. Most plaster subcontractors will also install stucco for exterior finishes.

Plaster subcontractors normally charge by the square foot of wall area covered or by the hour.

Cabinet Subcontractor: Manufactures and installs cabinets, shelving, bookcases and special trim either on the job or in a custom cabinet shop.

Note! Cabinet subcontractors build custom cabinets to meet specific project specifications. Many lines of good quality pre-manufactured cabinets can be found at building supply stores, kitchen and bath specialty shops and some lumberyards. Pre-manufactured cabinets are produced in stock sizes and may or may not fit precisely in the allotted spaces. Spacers or other methods of making the cabinets fit the spaces must then be employed. Since pre-manufactured cabinets are fabricated in a factory and shipped to a retail sales facility, the price you pay would only include the units themselves and shipping charges. Installation, if done by the seller, is extra. You may also elect to have your framing subcontractor do the work (at an extra fee) or install them yourself. On my own specific projects, I have always saved money by using cabinet subcontractors rather than purchasing pre-manufactured units.

Cabinet subcontractors normally charge by the foot (in length) of cabinet, shelf, bookcase, etc., by the hour plus material or by the job.

Paint Subcontractor: Applies paint and/or stain to all interior and exterior trim, windows and doors, paints all interior walls, exterior siding and other paintable components.

Paint subcontractors normally charge by the hour or by the job.

Wallpaper Subcontractor: Installs wallpaper on interior walls.

Note! Most paint subcontractors also install wallpaper. I have listed it as a separate trade specifically for clarification.

Wallpaper subcontractors normally charge by the roll of wallpaper or by the hour.

Tile Subcontractor (Also known as Tile Setter): Installs ceramic tile, quarry tile, marble and similar finishes on walls and floors, for tiled bathtub and shower areas, on counter tops and swimming pools.

Tile subcontractors normally charge by the square foot or by the hour.

Floor Covering Subcontractor: Installs carpet, wood, vinyl and other synthetic floor coverings.

Floor covering subcontractors normally charge by the square foot or square yard of material installed. In most cases, however, installation of the floor covering can be included as part of the overall sales price when you purchase the material.

SPECIAL SUBCONTRACTORS—Contractors who perform a specific job associated with the project, but whose tasks are specialized and can be performed by common subcontractors. Special subcontractors include but may not be limited to the following classifications.

Waterproofing Subcontractor: Installs waterproofing materials for walls enclosing basements or other useable areas that will be below grade.

Waterproofing subcontractors normally charge by the square foot or square yard of area covered or by the job.

Countertop Subcontractor: Installs countertops of specialized materials such as granite, cultured marble or other types of solid materials.

Countertop subcontractors normally charge by the square foot or by the job.

Glass Subcontractor: Installs fixed (non-operable) or special (frosted or stained) glass. Such glass is always special ordered or custom-made for the project.

Glass subcontractors normally charge by the square inch or square foot of glass or by the job.

Insulation Subcontractor: Installs insulation in the walls (usually fiberglass batts or foam plastic), in the attic or above the ceiling (usually blown in cellulose or mineral wool or fiberglass batts) and under the floor (usually fiberglass batts).

Note! I have included insulation subcontractors under the special subcontractor classification, because, for many years, it was common for the general contractor's crew to install insulation. With the emphasis on energy efficiency in today's housing, installation of insulation is becoming more specialized.

Insulation subcontractors normally charge by the square foot and material density, weight or thickness.

Security Systems Subcontractor: Installs smoke, fire and burglar alarms, and/or closed circuit television systems and all related wiring and devices.

Security systems subcontractors normally charge by the system, number and quality of devices (sensors, keypads, amplifiers, etc.).

Hardware Subcontractor: Installs doorknobs, latches and locksets on all exterior and interior doors and/or pulls and in some instances, hinges on all cabinet doors.

Note! Normally the general contractor installs door hardware and cabinet hardware is installed by the cabinet subcontractor. Hardware subcontractors, however, are fast, relatively inexpensive and will install the hardware just before occupancy. The use of hardware subcontractors is attractive to some general contractors because installation of hardware too early greatly expands the risk of damage or theft. At this point, however, I am not convinced that hardware subcontractors are worth the small, but nonetheless additional cost.

Hardware subcontractors normally charge by the number of devices installed, by the hour or by the job.

Solar Equipment Subcontractor: Installs equipment for solar heating and water heating systems.

Solar equipment subcontractors normally charge by the system and amount of equipment installed.

Water Well Subcontractor: Installs wells for potable water (in areas where water service is not available) and for lawn sprinklers or irrigation systems.

Water well subcontractors normally charge by the depth of the well or by the job.

Irrigation Subcontractor: Installs lawn sprinklers and irrigation systems.

Note! Quite often irrigation subcontractors also install water wells. This is not the fact in all cases, however, thus the separate listing.

Irrigation subcontractors normally charge per foot of pipe installed or by the job.

Elevator Subcontractor: Installs elevators, dumbwaiters, chair lifts and other specific lift equipment.

Elevator subcontractors normally charge by the job.

Seawall Subcontractor: Installs sea walls and bulkheads (to protect structures on ocean, river or lake front property).

Seawall subcontractors normally charge by the job.

Swimming Pool Subcontractor: Installs swimming pools and all related equipment.

Swimming pool subcontractors normally charge by the job.

Landscaping Subcontractor: Installs trees, shrubs and grass. In most cases the landscaping subcontractor will haul away debris and fine grade property as well as install berms, landscaping timbers and other yard-oriented components.

Landscaping subcontractors normally charge by the plant or tree for materials and by the hour for installation.

Cable TV Subcontractor: Installs all wiring, boxes and devices for cable TV service.

Note! In many cases, the TV Cable Company serving the community will also provide a prewiring service if notification is given and arrangements made during the early stages of construction.

Cable TV subcontractors normally charge a fixed fee per outlet.

Entertainment System Subcontractor: Installs all wiring, speakers and equipment for music, computer and intercom systems.

Entertainment system subcontractors normally charge by the system, number and quality of devices (speakers, outlets, etc.).

Telephone Subcontractor: Installs all wiring, boxes and devices for telephone service.

Note! In all cases that I know of, the Telephone Company serving the community will provide the prewiring service if notification is given and arrangements made during the early stages of construction.

Telephone subcontractors normally charge a fixed fee per outlet.

Pest Control Subcontractor: Provides soil and yard treatments for termites, ants and other pests. Check the pest history in the neighborhood of your home site to determine the types of control necessary and the frequency of retreatment.

Pest control subcontractors normally charge by the job or contract.

Garage Door Subcontractor: Installs garage doors, garage door openers and all related equipment.

Note! The general contractor's crew usually installs garage doors if a job is under contract. Since you will be performing the tasks of the general contractor and, if garage door subcontractors are nonexistent in your area, the garage door will have to be installed by one of the three following entities:

The framing subcontractor (garage door installation is not usually included in framing);

The supplier from whom you purchase your door (installation is always extra, but expertly performed);

Yourself (all doors have instructions packed with the materials). Allow yourself several hours per door.

Garage door subcontractors usually charge per door and per opener.

As you can see, the number and type of subcontractors on any project can differ greatly. All subcontractors have at least one thing in common, however, and that is that each and every one is a businessman. Since the success of meeting your goals depends on the subcontractors you choose, dealing with them in a businesslike manner is very critical. Responsibility for hiring, negotiating contracts, coordination of tasks, payment and "call backs" will all fall squarely on your shoulders. Although each one of these items could be a major effort individually, with a systematic approach, they are broken down into easy to manage tasks.

SELECTION—The most reliable method of determining which subcontractor will perform closest to your standards and still stay within budget, is by referral. Before selecting a subcontractor in any trade, check as many sources as you feel necessary to make a sensible decision about its ability to perform. A good rule of thumb for selection of subcontractors (after you have obtained several names) is to categorize them into three groups.

The first group will be known as the "A" list and, will consist of the subcontractors that are steadily employed by the ten best general contractors and/or homebuilders in your area. By virtue of the fact that they are employed by the best builders in the area, the "A" list subcontractors are usually well qualified, reliable and employ the same tradesmen on a steady basis.

The second group will be known as the "B" list and will consist of other qualified subcontractors that do work for any general contractor, homebuilder, individual or company. "B" list subcontractors are qualified to perform the required tasks, usually employ the same tradesmen (but tend to have a higher turnover rate among workers) and are reasonably reliable.

The third group will be known as the "C" list and, consists of subcontractors that have lower qualifications and do very little if any work for contractors, employ few, if any, steady tradesmen, (most workers are day or occasional laborers) and fall within the range of moderate to none in reliability. "C" list subcontractors tend to be sloppy, waste and/or steal materials, produce a low quality product and attempt to get paid in advance or for work which has not been performed.

In all cases your first choice should be from the "A" list. If you cannot hire a subcontractor from the "A" list, select from the "B" list. If you cannot get a subcontractor from the "B" list, start the listing process again. Do not, under any circumstances, hire a subcontractor from the "C" list!

You can establish leads for the listing process by visiting the job sites of the best builders in the area and speaking to the workers, observing the names on the subcontractors vehicles, observing the names on permit placards or asking for business cards. In addition, you can obtain leads from contractor salespersons at the various material supply stores and from the building official. The building official should not recommend one subcontractor over another, but will provide you with several names or verify the names you have collected, based on the quality of their work and, may even assist you in placing them on the "A", "B" or "C" lists.

You will certainly need to know if the person has the qualifications and experience to give you a quality job. By all means ask to see licenses (city, county, and state) and trade cards (Master or Journeyman) if issued by the governmental jurisdiction having authority in your area. Check with your local building inspection department and/or licensing agency to determine if the cards and licenses are current.

Ask for at least three recent job addresses or the names of three persons for who work (of the same magnitude as yours) was performed. Do not be afraid to check references on your subcontractors. Reliability, qualifications, quality of work and fair business practices should be your major concerns when considering subcontractors.

Note! In many cases, you can hire a subcontractor to supply the technical skills and perform the complicated tasks while you assist with common labor. The majority of the tasks in any construction trade do not require skilled laborers. If you are not afraid of work and have a true desire to save money, you can perform many of the common labor tasks yourself. You can also request a list of tasks from the subcontractor that you can perform in the evenings after work or on weekends. Bear in mind that many tasks in the construction trades are physically demanding, but each hour you work, you save the money that you would pay another person to perform that same task.

All subcontractors should have or be able to obtain the tools necessary to carry out all phases of the trade. Do not, under any circumstances, purchase or provide tools or tool replacements for a subcontractor. Doing so will only amount to an additional and unnecessary expense for you.

COORDINATION OF TASKS—Prepare a list of subcontractors that will be needed for your project. Next, prepare an order of trade progression for those subcontractors. Use the information given in the beginning of this chapter to establish your list.

PAYMENT—The method of payment to all subcontractors should be established early in negotiations. In all cases, the method should be in writing (preferably in contract form) and signed by the subcontractor and you. The early stages of this chapter explain the most common methods of payment for specific trades. Do not be afraid of alternate methods suggested by the subcontractor. Do, however, make sure you understand what service and material the subcontractor will provide for its services.

Pay all subcontractors and other persons associated with your job by check. Keep a payroll sheet for each trade. Document the number of workers on the job, the number of hours each person works, the amount of each check and the date each check is written. If a single check is given for a crew, information for each individual worker should nonetheless be documented.

CALL BACKS—After each trade is completed, inspect the work well and prepare a list of items (normally referred to as a punch list) that have not been completed, are not up to the job specifications or are not up to your standards. Require the subcontractor to bring all punch list items up to your satisfaction before final payment is made. Failure to do this may result in your having to live with the discrepancies or pay another person to correct them.

BARTERING—As mentioned before, subcontractors are businessmen. They will perform certain tasks for compensation. Compensation, however, does not always mean monetary payment. Use your ingenuity! You may have a product, service or profession that can be readily transformed into compensation. Bartering has been in fashion for centuries. There is not need to abandon its concept if you can make it work for you. In most cases, both parties in a bartering situation will receive more goods or services than a fixed fee or commission will produce. By putting your bartering as well as professional skills to work, you can realize some decent cost savings.

INTERNET—Although the Internet is a very useful and powerful tool, it is not my choice for selecting subcontractors. If the Internet is used strictly as a source for locating subcontractors in your area, however, it can be effective. I firmly believe that at least one initial face-to-face consultation with each subcontractor is a necessity. It is foolish to base your hiring of any tradesmen on a flashy homepage or clever Internet address. If you are in a rural area and believe the Internet will broaden your list of potential subcontractors, use it with caution. In any event, do not make your selection or enter into a contract until you have an actual meeting and review references.

Develop a checklist based on the following format, keep it handy and utilize it throughout all stages of construction.

TABLE 4

SUBCONTRACTOR CHECKLIST

SUBCONTRACTOR	NAME	PHONE #
Excavation	________________;	________________
Concrete	________________;	________________
Steel	________________;	________________
Masonry	________________;	________________
Framing	________________;	________________
Electrical	________________;	________________
Plumbing	________________;	________________
Mechanical	________________;	________________
Gas	________________;	________________
Roofing	________________;	________________
Dry Wall	________________;	________________
Plaster/Stucco	________________;	________________
Cabinet	________________;	________________
Paint	________________;	________________
Wallpaper	________________;	________________
Tile	________________;	________________
Floor Covering	________________;	________________
Glass	________________;	________________
Insulation	________________;	________________
Security Systems	________________;	________________
Solar Equipment	________________;	________________
Water Well	________________;	________________
Lawn Irrigation	________________;	________________
Elevator	________________;	________________
Seawall	________________;	________________
Swimming Pool	________________;	________________
Landscaping	________________;	________________
TV Cable	________________;	________________
Entertainment System	________________;	________________

Telephone   _______________________ ; __________________

Pest Control   _______________________ ; __________________

Garage Door   _______________________ ; __________________

Other   _______________________ ; __________________

                   _______________________ ; __________________

                   _______________________ ; __________________

                   _______________________ ; __________________

                   _______________________ ; __________________

                   _______________________ ; __________________

                   _______________________ ; __________________

CHAPTER SIX

CONTRACTS

A contract is nothing more than a formal agreement (usually written) between two parties. Although verbal contracts are common in the construction trades, I strongly suggest that you make all of your agreements and arrangements in writing. Written contracts protect both the subcontractor and yourself from misunderstandings that could and usually do develop.

Most subcontractors submit written proposals or "bids" on forms that double as contracts. Such forms are usually well prepared and contain all the necessary information required for task performance and completion. Read the proposal carefully and compare it with the information and items covered in this chapter. If you are not fully satisfied with the format or contents or if the subcontractor does not have such forms, you should prepare your own contract.

Contract documents should be kept as short as possible. Specific information, however, is necessary for a complete agreement. Any contract you prepare should contain at least the following information:

The names of the persons, companies, corporations or partnerships involved;

The date the agreement is made;

The location of the project;

A description of the work to be performed, including all materials supplied by the subcontractor;

General conditions;

The type and amount of compensation agreed upon;

The method of payment; and,

A reasonable time period in which the work is to be completed.

You and a responsible representative of the subcontractor must sign all contracts. If the subcontractor prepares a contract, make sure you fully understand all the terms before you sign it. If any part of a contract is confusing or not clear, consult an attorney for clarification. Once your name is on the proverbial dotted line, you have a legal and binding document that under most conditions will hold up in any court of law.

The description of work to be performed should be as complete as possible. If certain materials are to be installed, be sure to include the type of material, brand name and the amount of material to be used.

FOR EXAMPLE: Do not specify; "paint, exterior of house, colonial yellow with autumn brown trim."

Do specify; "paint, exterior siding, two coats of ABC brand colonial yellow latex flat. All trim, to receive two coats ABC brand autumn brown latex semi-gloss. All wood surfaces to receive one coat ABC brand latex primer before paint."

Do not specify; "provide all materials and install roof covering."

Do specify; "install 240 pound ABC brand rustic brown, seal tab fiberglass shingles over 15 pound felt. Install aluminum flashing around chimney and in all valleys. Install pipe flashing and seal around all other roof penetrations."

You can obtain descriptions of the materials and lists of the trade components that will be necessary to perform the tasks from the subcontractors or suppliers. You may, however, elect to shop for materials yourself. In that situation, you can eliminate brand names, and state instead that the materials are to be supplied by you, the owner.

Note! Shopping for materials yourself can produce some real cost savings. Normally, subcontractors will purchase most, if not all of the required materials, from one supplier. Check various sources and purchase from the supplier that gives the best buy of the day. Before shopping, however, check your subcontractor's material costs. A trade discount offered to the subcontractor may already be the lowest possible price.

The general conditions section of the contract should contain information that pertains to all subcontractors collectively and have a general nature. You may wish to include other items, but in all cases include the following statements.

The subcontractor shall provide all tools and equipment necessary for the execution and completion of the work.

The subcontractor shall meet all licensing requirements of the jurisdiction having authority.

The subcontractor shall secure all permits and pay all fees involved before work is started.

Note! Some contractors charge a small fee for obtaining permits. You can omit that item and associated cost if you elect to obtain the permits yourself.

The subcontractor shall call for all required inspections, pay all inspection or reinspection fees and bear full responsibility for correcting discrepancies or code violations at its expense.

The subcontractor shall be responsible for employing persons qualified to perform in a skilled, workmanlike manner and shall not allow any employee to work while intoxicated or in any other manner physically or mentally unfit.

The subcontractor shall be responsible for cleaning the job site of all debris, residue or waste material generated by its trade or employees before the end of each workday.

The subcontractor shall be responsible for and shall provide, any and all insurance required by law or to protect the owner from any claim for loss of property or injury of any type including death, which may be the result of its trade or employees.

All work shall be performed in strict compliance with the (building code adopted by your local jurisdiction) building code and any other regulatory documents as adopted and used by (the local jurisdiction).

Note! Obtain the name and edition of the code governing the subtrade and the jurisdiction in which it is used, from the building official in your area.

Any changes to this contract or to the description of work stipulated herein, must be done so in writing with all changes detailed and the cost adjustments (either up or down) so stated. Such additions or deletions to this document shall be dated and signed by the owner and the subcontractor.

The subcontractor shall take whatever steps necessary to protect its work from damage. The subcontractor shall also take whatever steps necessary to protect other components of other trades from damage caused by the performance of any phase of its work.

The subcontractor shall be responsible for correcting any defective work or replacing any defective material at its cost, for a period of one year after the date of completion or within any time period longer than one year as may be prescribed by law.

Note! Some states require contractors to guarantee their work for a specified time period whether designated by contract or not. Be sure you read and thoroughly understand the warranty conditions of the materials and components before installation. Installation of any material or component, which is not in strict compliance with the manufacturer's instructions or recommended procedures, may invalidate the warranty.

The method of payment should be established for each subcontractor based on the amount of the contract and the length of time required for completion. In most cases, a request for payment (usually referred to as a draw) will be made when a certain percentage (usually 50 percent) of the work is completed. In some cases the subcontractor will not request payment until the job is completed. In either case, however, before payment is made, obtain certification that all required inspections have been performed and approved by the local building inspection department.

Note! Building inspectors look for violations of building codes and laws or ordinances they are charged to enforce only. Although the work should conform to local standards for workmanship, building inspectors do not, as a rule, perform quality control inspections.

Inspect the work yourself and make sure, especially on finish work that the job is to your satisfaction. Require all discrepancies or flaws to be corrected before payment is made. Be reasonable, however, in your inspection and expectations. Subcontractors are human and even in the finest homes that money can produce, flaws can be found. Your goal should be to produce the best quality home for the price you can afford to pay.

If more than one payment is required for any subcontractor, hold back at least 10 percent but not more than 15 percent of the requested amount of the draw. The provision for holding a portion of the payment should be included in the contract as a "retention clause" in the method of payment section. The compensation retained should be held until the job is completed and then paid after all obligations on the subcontractor's part have been fulfilled. If at any time the subcontractor fails to live up to the contract specifications, the fees retained, in addition to the balance owed, can be used to hire another subcontractor to complete or correct the work.

Keep the subcontractor's needs in mind, as well as your own, when a request for a draw is received. Make payment as soon as practical after receipt of the draw request. Remember that the subcontractor has employees, operating expenses and material costs that must be paid. The quickest way to develop bad relations on your project is to withhold payment on a draw without good and valid reasons.

An acceptable time period in which the work is to be completed should be stipulated in the contract. This information can be obtained by questioning the subcontractor. Again, be reasonable. Although under optimum conditions some subcontractors may be able to complete their respective tasks in as little as one day, additional time may be required for a variety of reasons. Do not be unrealistic. Based on the reason, you should consider allowing the subcontractor more time than agreed upon, as long as the delay does not affect other trades. If a situation develops that is beyond the control of the subcontractor, you should be prepared and willing to accept it without retribution.

It is common for subcontractors to take on more work than they can reasonably handle. This practice is justifiable, at least in the minds of the subcontractors, because of the "feast or famine" nature of the construction business as a whole. The thought of turning business away is so repulsive that the subcontractor will attempt to work it into it's schedule with the other jobs it may have going at the time. The result is normally a delay in time for all the jobs or hasty, haphazard work. The time period, therefore, is used as a contract tool to discourage the subcontractor from starting the job and then not returning to complete it within a reasonable amount of time.

The following document is a completed example of a contract containing all of the criteria listed above. It may be reproduced and used as is or it may be used as a base for a more complex document if one is needed. The words that are shown in italics correspond with the information that you will supply.

This contract is hereby agreed upon and made on the *14th day of March, 2009*, between *ABC Plumbing Co.*, hereafter referred to as the subcontractor, and *John T. Smith*, hereafter referred to as the owner, for work to be performed as specified below, on property situated at *Lot 10, Block 10, Springdale Subdivision in the City of Franklin Heights.*

For compensation of *Seven Thousand Two Hundred and Fifty Dollars ($7,250.00)* the subcontractor agrees to furnish *all materials, fixtures and labor for the installation of the plumbing, including the drain, waste, ventilation and potable water supply systems for the new residence situated at the location stated above.*

Compensation by the owner to the subcontractor shall be made in the following manner. *One third (1/3) of the total contract amount ($2,416.67) less 15 percent retainage ($362.50) for a payment of $2,054.17, after all piping has been installed, tested and approved by the Franklin Heights Building Inspection Department; one third (1/3) of the total contract amount of ($2,416.67) less 15 percent retainage ($362.50) for a payment of $2,054.17 after all fixtures have been installed, tested and approved; one third (1/3) of the total contract amount of ($2,416.67) less 15 percent retainage ($362.50) for a payment of $2,054.17 after the sewer connection has been completed; and, $1,086.00 (retainage) after all "punch list" items have been completed.*

Completion of all work performed by the subcontractor shall be accomplished within *ninety (90)* working days following the date this document is signed. If the work specified is not completed within the time period stipulated above or, if the subcontractor fails to meet any of the terms of this contract, the owner has the right, after written notification to the subcontractor, to have the work completed by whatever means the owner deems necessary and may deduct the cost or value of such work from compensation due the subcontractor by this document. If the compensation required to complete the work exceeds the unpaid balance due to the subcontractor, the subcontractor shall reimburse the owner for the amount or value of the difference.

General Conditions:

1. The subcontractor shall provide all tools and equipment necessary for the execution and completion of the work.

2. The subcontractor shall meet all licensing requirements of *the City of Franklin Heights.*

3. The subcontractor shall be responsible for securing all permits and for payment of all fees involved before work is started.

4. The subcontractor shall make arrangements for all inspections required for its work, pay all inspection and reinspection fees and bear full responsibility for correcting discrepancies or code violations.

5. The subcontractor shall be responsible for employing persons qualified to perform with a reasonable amount of skill and workmanship and, shall not allow any employee to be on the job site while intoxicated or in any other manner physically or mentally unfit to perform the required work tasks.

6. Before the end of each workday, the subcontractor shall be responsible for cleaning the job site of all debris, residue or other waste material generated by its trade or employees.

7. The subcontractor shall provide any and all insurance required by law, including workman's compensation, in an amount sufficient to protect the owner from any claim for loss of property or injury of any type, including death, which may be the result of the subcontractor's trade or employees.

8. All work shall be performed in strict compliance with the *Plumbing* Code and any other regulatory document as adopted by *the City of Franklin Heights.* If *the City of Franklin Heights* had not adopted a *Plumbing* Code, all work must conform with commonly accepted industry standards.

9. Any and all amendments to this contract or to the description of work stipulated herein must be done so in writing, with all amendments detailed and the cost adjustments, either up or down, so stated. Any and all amendments to this contract shall be dated and signed by the owner and the subcontractor.

10. The subcontractor shall take whatever steps necessary to protect its work from damage. The subcontractor shall also take whatever steps necessary to protect other components of other trades from damage caused by the performance of any phase of its work.

11. The subcontractor shall be responsible for correcting any and all defective work or replacing any and all defective material, at the subcontractor's expense, for a period of one year after the date of completion or within any time period longer than one year as may be stated by warranty or prescribed by law.

12. Other Provisions____
    (Include 10 to 15 lines to document any other provisions that are mutually agreed upon.)

13. Signed and Sealed:

_________________________   _________________________

John T. Smith, Owner   David E. Jones, President
      ABC Plumbing Company

CHAPTER SEVEN

OBTAINING PERMITS

Before work of any kind is started on your project, be sure you have obtained all of the required permits. In some areas, a single building permit may be all that is required, whereas in other areas a variety of permits may have to be obtained. In this chapter we will explore the different types of permits and the various utility connections that may be required, the procedures for obtaining them and the agencies that will have jurisdiction.

The various types of permits that may be required have already been mentioned in Chapter Three. For convenience, however, and to stress their importance, they will be listed again here by trade and in more detail.

Tree Removal/Lot Clearing Permit: Allows the removal of protected trees or clearing of the property.

Grading Permit: Allows the alteration of the natural grade or contours of the property.

Building Permit: Allows the construction of the foundation and superstructure (walls, floors and roof) including finishes and trims.

Plumbing Permit: Allows the installation of all drain, waste and vent piping, as well as all potable water supply piping, water heater(s) and plumbing fixtures.

Electrical Permit: Allows the installation of all wiring, electrical equipment, panels and subpanels, switches, receptacles, light fixtures and electrical devices.

Mechanical Permit: Allows the installation of all ductwork and heating, ventilating and air-conditioning appliances.

Gas Permit: Allows the installation and connection of gas piping to gas appliances.

Septic Tank Permit: Allows the installation of a septic tank and drain field lines (where public sewage disposal systems are not available.

Note! Septic tank permits usually require the property to be subjected to a percolation test to determine the absorption capabilities of the soil. An engineer or other competent person must perform percolation tests. In some areas, an employee of the Department of Health will perform the test at little or no charge.

Seawall or Bulkhead Permits: Allows the construction of sea walls or bulkheads on water front property and, in some areas, private fishing or recreation piers.

Swimming Pool Permit: Allows the installation of a swimming pool and all of its related equipment.

Encroachment Permit: Allows encroachment on public property (usually for driveway connections with a public street, also known as a "curb cut").

The building, plumbing, electrical, mechanical, gas, grading, swimming pool and seawall/bulkhead permits are normally issued by the building official or a person within the building inspection department. If your building site is located in a jurisdiction that does not employ a building official, the city/county clerk or the fire chief may issue the permits.

Septic tank permits are normally issued by the local or state Department of Health. In some cases, however, the building or plumbing official may issue septic tank permits.

Tree removal/lot clearing permits are normally issued by the city or county planner or a person within the Planning Department, the zoning administrator or a person within the Zoning Department or the building official.

Encroachment permits are normally issued by the Street or Highway Department having jurisdiction over the public right of way adjoining your property.

A water service connection or "water tap" will be required if the neighborhood in which you are planning to build your house has public water service available. The actual water tap is nothing more than the installation of a water meter connecting your private potable water supply piping with the public water system. As compensation for the meter, equipment and labor for installation, a tap fee will be charged. The tap fee (usually based on meter size) must be paid in advance at the administrative offices of your local Water or Public Works Department. In addition to a water tap, a sewer connection or "sewer tap" will also be required where sewer service is available. The sewer tap consists of a physical connection of the building's sewer with the public sewage collection system. Both water and sewer tap fees should be paid at the same time. In some jurisdictions the payment of water and sewer tap fees can be paid at the same location as well, but do not be surprised if you must travel to two different entities or locations to pay the fees.

Note! Verification of the location and depth below grade of the sewer tap is available from the Water and Sewer or Public Works Department. It is advisable to locate the tap and note its depth below grade before the first floor level is established. The location of the tap is recommended to allow for the proper fall or slope of the sewer pipe from the house to the public sewage system. Failure to obtain the proper fall will result in constant clogging of the sewage system.

When making an application for any permit, you can eliminate unnecessary delays and make the process relatively painless if you keep one thought in mind; be prepared! Make sure that all of your documentation is up to date and accurate. Apply the "Systematic Approach" by obtaining all necessary applications and permit procedure information well in advance. If no specific process exists in your area, the following procedure should be used as a guide.

Variances: If a zoning variance is necessary for construction, apply for it as early in your planning stages as possible. A variance may take several weeks or longer to process. You also face the possibility of having the variance request rejected, which would require, at minimum, a building modification or conceivably a complete redesign.

Aesthetic Review: You may be required to submit your building plans for approval by a community appearance or architectural review board. Such boards may be established by governmental jurisdictions as well as private subdivisions. The board will be primarily concerned with the exterior appearance of your building, therefore, a detailed set of elevations is a must. Since elevations are two dimensional, floor plans should also be provided to allow the people that are reviewing the design to be able to distinguish any projections or recesses from flat wall surfaces. In addition, some boards may require sample materials such as brick, roof shingles and exterior paint colors to be submitted with the plans.

If you are required to have an aesthetic review before construction, submit a complete set of plans and material samples as soon as you obtain them. If at all possible, be present or have someone represent you at the meeting.

Soil Test: If your proposed building site has questionable or unusual soil conditions, a soil test will be necessary. An engineer or qualified soils expert can only perform a soil test. A properly performed soil analysis takes time, therefore, be sure to order it early and have the results in hand before making application for the building permit. The building official may require one copy of the test results to be filed and remain with your building permit application. Make sure you keep at least one copy with your building permit receipt for future reference.

Building Permit Application: Before a building permit can be issued, an application form must be completed and submitted to the permit issuing authority. Specific information is then transferred from the application to the permit. It is critical, therefore, to enter only accurate information on the application form. Obtain application forms from your local Building Inspection Department or administrative clerk's office well in advance of the anticipated start of construction. Take the applications home and study them along with any accompanying instructions. After you are sure that you understand what information is required, complete the application. As you work on the application don't forget to type or print legibly in ink, fill in all of the spaces, provide all additional information requested and sign in the space provided.

Most building permit application forms are relatively easy to complete. For specific information, however, you may need to refer to other documents. When filling out your application, it will be much easier if you have a set of plans, the plot plan and your property title handy.

Specific applications may request a variety of information, but usually ask for at least the following.

A legal description of the property (usually a lot, block and section or subdivision). An accurate legal description will be found on your property title or on your plot plan.

Owner of property. The owner(s) of the property must be as designated on the deed or title for the property.

Zoning classification. Zoning information can be obtained from your local building official or zoning administrator.

Recording of title. Information concerning the recording of the property title (a book number and page number designating where the instrument may be located in the public records) is contained on the title itself, usually in typewritten or rubber stamp form.

Name of the Contractor. On the contractor identification line, designate self or owner.

Total valuation or cost of the work. The project valuation amount should reflect the total cost of construction, including all subtrades. Be sure that the construction valuation is realistic, but do not include costs for the property itself, landscaping, driveways, walks, patios etc.

Note! Some jurisdictions will not issue a building permit if the value shown on the application is lower than a base minimum established for the area. For example: a base minimum of $75.00 per square foot of building area may be established in your particular area. If your total estimated construction cost amounts to $68.00 per square foot, your project will nonetheless be valued at the higher rate. The reason is to maintain property values at a stable and consistent level for tax and resale purposes. Your actual construction cost and the overall value of the completed project may be two different amounts. Bear in mind that the reason that you are reading this book in the first place is to gain the highest value house for the lowest possible cost. Check with your local building official to determine if a base minimum evaluation exists in your area.

General information, such as total building area in square feet (heated and unheated), number of stories, number of rooms etc. may also be required to be entered on the application. Such information can be found on your building plans or specifications.

Upon completion of the application, submit it along with your building plans, specifications, plot plan (more than one copy of each may be required by your jurisdiction, determine the number in advance) and a check for the appropriate fee to the building official or permit issuing authority.

Note! Remember that the plan review and permit processing takes time. You can assist the building official tremendously and expedite the issuance of your permit by presenting accurate, clear and detailed plans and specifications. If, during the building permit application process, you run into problems or a question arises that you can't answer, consult your building official.

In general, the permit application process is not complex or deeply involved. In fact, in most areas, the process is relatively simple. If the process in your area is simple, you may be tempted not to develop a checklist at all. Prepare one anyway. Although your checklist may be very short, it will help you maintain a complete record of your construction project.

Develop a checklist for the building permit application based only on what is required by your jurisdiction. The following text contains all the elements necessary to develop a comprehensive permit application checklist. For simplicity and ease of use, you may prefer to eliminate the elements that do not apply to your specific property. The words that are shown in Italics correspond with the information that you may include on your checklist.

Your checklist should contain all conceivable permits that you will obtain during the entire course of your project. Start with the heading Permits Required and list all preconstruction permits including the

                                    GARY F. WIGGINS

tree removal permit or *clearing permit, excavating permit and grading permit.* Go on to include all construction permits such as the *Building, Plumbing, Electrical, Mechanical, Gas, Septic Tank, Swimming Pool* and *Seawall/Bulkhead* permits. Also include *Encroachment/Curb Cut* permits and a line or two for any other unanticipated permits.

In many cases, your permits will be tied in with other approvals, so include them on your permit application checklist. Other approvals may include required taps. If so, provide spaces for w*ater* and *sewer taps,* and the d*epth of sewer tap* (below grade).

Various reviews may also be required before you are permitted to start construction, so include on your checklist, spaces for any *variance required* by the Zoning Board of Adjustments, for your jurisdiction. State the *yard or yards* that require a variance and the *amount of variance* necessary. Also include the *meeting date, time, place* and *result* of your request.

If you are required to have your plans approved by a Community Appearance or Architectural Review Board, provide spaces for the *meeting date, time,* and the *result.*

In most cases the building official will perform a plan review. Therefore, you should include spaces for the *date submitted,* any *changes required,* the *date resubmitted* and the *date* the plans are *approved.*

If tests are required before your building permit can be issued, include a section with *soil test, required* the *date performed,* and the *result.* Provide spaces for *percolation test required,* the *date performed* and the *result.*

Finally, you should have a section for general application information, including the *lot number, block number* and *section or subdivision.* Also include the *owner as shown on the title,* the *book number* and *page number* where the title is recorded and the *zoning district.* Finish your checklist with general information about the house such as the *total anticipated cost, total building valuation, total heated area, total unheated area, total garage area, total area* of all *decks, porches* and *balconies,* the *number of stories, number of rooms, number of bathrooms* and *number of bedrooms.* Don't forget to include a line or two for *other information.*

CHAPTER EIGHT

SITE PREPARATION

As soon as the building permit is issued, you are ready to start your project. Actual construction cannot start, however, until the site is properly prepared. This chapter will address site preparation and the tasks involved. If, during the process, you discover that one or more of the specified tasks may not be necessary, simply eliminate those steps. To expedite the site preparation process, perform all applicable tasks in the sequence or "Order of Progression" given.

Posting the Permit: The very first task in the order of progression is the posting of your building permit. The building permit placard should be tacked to a piece of plywood, which will form a rigid backing that will allow easy attachment to a post or stake driven into the ground. The permit placard should be posted at the front of the building site so as to be clearly visible from the street. The area around the placard should be clear of obstacles to allow accessibility for the various inspectors.

Although the permit placard is normally very durable, care should be exercised in protecting it from the weather. Most building supply stores stock weatherproof permit placard display cases that are free standing or can be mounted on the temporary service pole. If at all possible, avoid posting your building permit placard on a tree.

Survey: It is your responsibility and obligation to obtain a survey of the property before any work is started. The property line must never be assumed. Make sure that all corners of the building site are marked and clearly visible throughout construction. Care must be exercised to preserve the property corner markers in the exact position that the surveyor initially places them. If the corners happen to get damaged, lost or disturbed at any time, have them relocated as soon as possible. Construction string lines stretched from corner to corner will establish the lot boundaries.

Note! Do not, under any circumstances, disturb or allow any of your subcontractors to disturb any property, shrubbery, trees or items of any kind outside of your property boundaries. Do not allow workers to use adjoining property as a toilet or to deposit any waste or trash. Do not allow materials to be delivered to or stored on adjacent property or allow delivery or construction vehicles to cross any portion of adjacent property without written permission from the property owner. If permission to use or store materials on an adjoining lot is granted by the property owner, be sure to repair or replace any damaged items and return the grade to its normal condition upon completion of your project. Damage to any item on an adjoining property by you or any of your subcontractors or suppliers may result in a civil action (law suit) being taken against you.

Clearing: Once your property lines are established, the site is ready to be cleared. Identify all trees that are to remain on the site by tying colored ribbons (available in rolls from your building materials supplier) around the trunks, approximately five feet above grade. Remove only those trees that are poorly developed, diseased or dead or that will be in the way of development.

Grade Alteration: Grade alteration can be performed as soon as the site is cleared. If the lot is low and requires filling, make sure the fill material is clean and free of debris. If the lot must be raised a considerable amount, and since the foundation must be installed in natural or undisturbed soil, you can save yourself some additional work by filling after the foundation, basement or foundation walls, pilasters and piers are in place.

Toilet Facilities: Proper and adequate toilet facilities must be established on site before workers start on full workday schedules. The most common method for providing a job site toilet is to rent one. Portable chemical units may be rented on a monthly basis. Such rental units are usually well maintained and cleaned and serviced by the rental company on a periodical but frequent basis.

Portable chemical toilets are rented by construction equipment or specialized (toilet) rental businesses. Since rental is on a monthly basis, and to assure delivery when you need it, you should locate a rental business and reserve a unit several weeks in advance. Before renting a unit, however, make sure that the company (and the unit) is approved by your local Health Department.

Insurance: You, as owner of the property on which the construction is taking place, are ultimately responsible for personal injury that may occur on site. An insurance policy (known as "builders risk") with adequate coverage should be purchased in advance of the start of construction. Along with personal injury, you should also purchase coverage for fire and theft of materials, tools and equipment. Consult with your insurance agent about the various types of policies and coverage that is currently available for construction sites and projects.

Electric Service: From this point forward, your subcontractors will be using a variety of power tools. As soon as your grade is established, call your electrical subcontractor and have it install a temporary service pole.

The temporary service (also known as saw service) will require an inspection by your local Building Inspection Department before the power company will install a meter. Your electrical subcontractor will call for the temporary service inspection upon completion of installation.

A deposit for the meter and/or service will be required by the power company before electrical current will be supplied. In order to have your electrical service at the time your workers are ready to start, make application and pay the deposit at your local power company business office at least two weeks in advance of your anticipated starting date.

House Location: Location of the house on your lot is accomplished by the setting of batter boards. If you are not experienced in setting batter boards, call in your framing subcontractor. Once the batter boards are in place, construction string is stretched between marks, nails or notches on the boards to form the outside perimeter of the building. At this stage, the layout on the site should match the plot plan (which is used to locate the house on the site).

NOTE! Quite often, foundation subcontractors will install batter boards.

The string lines also establish the elevation of the first floor above the exterior grade. If you are building in an area that is subject to flooding or if the first floor elevation must be documented or certified for any other reason, a land surveyor or engineer will be required to make the certification.

Water Service: From this stage forward, water will be necessary for mixing mortar and other materials and for clean up purposes. Pay the water and sewer tap fees about one week in advance of the start of construction to allow for adequate scheduling time for installation of the meter. After the meter is installed, call your plumbing subcontractor and request the installation of a temporary water spigot close to the meter.

Note! It is most common to install the building water and sewer service piping at the very end of the project. I prefer, however, to not only install the building water and sewer service piping, but also any below grade gas piping, conduit or direct burial cable for electric, telephone and cable television service at the very beginning of the project. Doing so allows earlier usage of the plumbing system, as well as a settling period for the back filled soil, reducing the need to refill trenched areas after the final grading is complete. If you elect to install the items mentioned above during the early stages of construction, notify the plumbing, gas and electrical inspectors in advance to coordinate the various inspections that may be needed.

Ground Condition: Check the entire area around the perimeter of the house for low or soft spots. Low spots will hold water after a rain and create sloppy, muddy conditions. The resulting mud tends to be very slippery and can be the cause of severe accidents. Soft spots may give way under the load of heavy equipment or trucks delivering goods and materials to the site. In such cases, precious time and manpower must be wasted in freeing the bogged down equipment. In extreme cases, a tow truck may be required to pull the stuck vehicle from the soft area, which can cost you time and good money without returning any value to the property.

If areas exist on site that are low, wet or soft, keep all trucks, equipment and stored materials away from them until the situation can be corrected. If water is standing anywhere on site, excavate temporary drainage ditches to allow for the removal of the water and drying of the area.

Trash Removal: Designate an area on the site for a trash receptacle to be constructed or installed by a trash removal company. Make sure the area is close to the street, so it may be serviced by the appropriate trash collection agency or company. Make arrangements for trash removal about one week in advance, to allow for scheduling or possible route changes that may have to be made.

Some construction materials (brick, block, concrete, sheetrock, etc.) may not be accepted for removal because of weight or inability to be compacted. Arrange for removal of such items separately. You can even consider hauling the debris to the landfill or dump area yourself, if you have access to a truck. All trash generated by subcontractors should be removed by them or deposited in the receptacle at the end of each workday.

Your trash receptacle should be large enough to contain various sized boxes and material wrappings. It should also be able to contain small, light waste materials that may have a tendency to blow onto adjoining lots if a wind develops. It will be up to you to require and remind all workers to deposit used food containers and wrappers, as well as other personal trash in the receptacle on a daily basis.

Trash Burning: Some local jurisdictions will allow construction trash to be burned. With air pollution and the danger of fires getting out of control, however, most cities and counties have laws preventing burning of any kind. In all cases before you attempt to burn any trash on site, check with tour local Fire Department. In local jurisdictions where fires are allowed, they are usually regulated and require burning permits. Even if a burning permit is not required, however, make sure the Fire Department has your name, location of property and the dates and times you plan to burn. Monitor all fires carefully, do not allow them to burn unattended and by no means after you or your workers have left the site for the day. After the fire is completely burned out, soak the area well with water. Notify the Fire Department when the burning period has been completed and the fire is completely out.

The order of progression for site preparation is now complete. Develop a site preparation checklist to establish your specific preparation needs and to maintain the proper sequence of events. The following text contains all the elements necessary to develop a comprehensive permit application checklist. For simplicity and ease of use, you may prefer to eliminate the elements that do not apply to your specific property. The words that are shown in Italics correspond with the information that you may include on your checklist. To maintain a time schedule for future reference, include a space after each item for the entry of a date after each task is completed.

The site preparation checklist should include all the following items. The *building permit posted, survey obtained, corners visible, site cleared, trees marked, grade lowered, grade filled, toilet facilities provided, health department approval, insurance obtained, electric meter deposit paid, temporary service pole installed, temporary service pole inspected and approved electric meter installed, batter boards installed, certified grade elevation* (for flood purposes), *water tap and meter deposit paid, water meter installed, temporary spigot installed, sewer tap fee paid, sewer tap made, sewer tap inspected and approved, water piping installed, water piping inspected and approved, sewer piping installed, sewer piping inspected and approved, gas piping installed, gas piping inspected and approved, conduit installed for electric, telephone, cable tv, cable installed for electric, telephone, cable tv, electrical conduit and/or cable inspected and approved, ground condition checked, trash receptacle installed, trash removal arranged, trash burning allowed/permit issued,* and *fire department notified of burning date(s).*

CHAPTER NINE

CONSTRUCTION

Actual construction is a relatively simple undertaking if all of the individual tasks are performed in proper sequence. That proper sequence is what I have been referring to as the "order of progression."

This chapter is going to take you on a journey through an entire order of progression for a typical single-family residence. The information and techniques explained in this chapter will apply to the vast majority of houses that can be constructed. If your house is very unusual or extremely complex in design, the order of progression may have to be altered, but will still be applicable.

Never forget the fact that you are the contractor and in complete charge and control of the project at all times. Be on the job and supervise as much of the work as possible. If you can not be on site the same time as the workers, make a list of tasks to be performed and set your goals for each workday the night before. Review your list each night and evaluate the progress made on a daily basis before preparing a new list for the following day. Set realistic daily goals, but also be prepared to make adjustments when necessary. If for some reason you do not meet the specific goals you set for a day do not panic, it certainly is not the end of the world. Unless you lose a day for bad weather or lose work time for any other reason, you can incorporate any incomplete goals into the next day's schedule.

Establish your task schedule on paper to give a visual course to travel. Once established, make every effort to stay on schedule. One of the best methods of staying on schedule and avoiding unnecessary expenses is to stick to your plans. Changes to your plans after construction has started are very costly. That one point cannot be stressed too often or too much.

Since you will be responsible for ordering most of the materials, calculate the quantities and order well enough in advance to assure delivery by the day you need it. If the required materials are not on the job when the subcontractors are ready to start work, unnecessary delays will result.

ORDER OF PROGRESSION—Excavation: After the site is cleared and if your house contains a full or partial basement, excavation should begin. In the majority of cases, excavation can be accomplished with a tractor. A tractor will remove almost all of the earth in a quick and efficient manner. In addition, a tractor will remove large rocks and tree stumps. Hand excavation, which costs far less per hour, takes much more time to accomplish. Some hand excavation (trimming) will be necessary, however, so plan for both.

Note! After the lot is cleared and prior to the foundation excavation, batter boards must be installed. The batter boards may be installed by the foundation or framing subcontractor.

Foundation: The building foundation is the very first component to be installed. Since the weight of all other materials bear directly upon the foundation system, it is the only component that cannot be readily repaired or replaced. Make sure that all foundation trenches and/or pier footings are excavated in strict compliance with your plans and specifications. Do not undersize the trenches or footings in either width or depth. Under sizing may save a few dollars in concrete costs, but may create major headaches if structural failure occurs.

If your foundation plan specifies steel reinforcing, make sure it is installed straight and tied tightly at each support and where overlaps occur. Grade stakes (the stakes that are placed in the bottom of the excavation to designate the depth to which the concrete is to be poured) should be short sections of reinforcing steel. Never use or allow the use of wooden grade stakes in a foundation.

Your excavation subcontractor, if used (in some cases the framing subcontractor will also prepare the foundation), will perform the task of excavating the trenches. If your excavation subcontractor does not install reinforcing steel, you will have to call in your steel subcontractor. Excavation and installation of reinforcing steel for a foundation without a basement can be performed almost simultaneously.

After installation of the reinforcing steel is complete, perform a visual check of the foundation. Make sure all excavations go into undisturbed or natural soil and, are free of roots, stones, water and trash.

Once you are convinced that the foundation is ready to be filled with concrete, call for your foundation inspection. Try to be on the job site when the building inspector arrives in case any questions need to be answered or discrepancies corrected. The building inspector will check the width and depth of the excavation, the bracing and tying of your reinforcing steel (if installed) and the thickness of the actual foundation (by measuring the height of the grade stakes). The inspector will also verify the soil conditions and check for nonapproved items (roots, stones, loose soil, trash, water, etc.) in the trenches.

Note! If your house is to be built on a monolithic foundation (footings and basement or first floor poured together, forming a single component system), your plumbing drain lines, water supply piping, gas piping and any conduit or below floor mechanical ventilation system must be installed and inspected before or at the time of the foundation inspection. In addition, your pest control subcontractor should treat the area of the slab before the vapor barrier is installed. The vapor barrier and concrete reinforcing wire mesh for the slab must be in place before the monolithic floor and foundation inspection can be made. After the building inspector inspects and approves the foundation system, you may order your concrete.

Placing concrete in foundation trenches is not a technically difficult task. You can do it yourself with the assistance of two or three helpers or you can hire three common laborers. If you decide to do it yourself, however, be prepared for a workout.

For the best possible foundation, do not dilute the concrete with water. Never place concrete just before or during a rainfall. Make sure that all concrete is placed at truck consistency to the tops of the grade stakes. Level and allow the concrete to thoroughly cure before applying heavy or uneven loads.

With good site and weather conditions, the process of foundation installation should be performed in one day, but in some cases, may take two days. Unless you have unusual conditions, severe problems or a very complex foundation design, you will be wasting time if you spend more than three days on the process.

Masonry Foundation Walls And Piers: In most instances, the masonry piers and exterior walls, from the foundation to the first floor level (also referred to as curtain walls), may be installed the day after the concrete is poured for the footings. Give your masonry subcontractor about five working days notice to fit you into its schedule.

Order the block, mortar and sand and, request delivery late on the date you plan to place concrete for the foundation system. I request my suppliers to schedule my projects for the last delivery of the day. Late delivery will assure that the materials will not be in the way of concrete trucks as they are unloading. It will also allow the delivery truck drivers to place the materials in specific designated areas around and closer to the foundation.

If your first floor is to be concrete, make sure you specify header blocks (blocks manufactured with a four inch by four inch section removed length wise, to receive concrete) for the top course of your curtain wall. If your first floor structural system is to be constructed from wood, a four-inch solid concrete or masonry cap must be provided on top of the final course of masonry.

Note! My personal preference is to fill all foundation walls and piers with concrete, to form a solid structural wall instead of a hollow block wall between the footings and the first floor, regardless of the height of the wall. When a basement is specified by plan or a high foundation wall and pier system required by design or site necessity, however, the building code may require filled blocks. If filled blocks are required, make sure they are done so in strict compliance with the building code and good engineering practice.

Unless your house is very large, complex or has a basement, the installation of masonry foundation walls and piers should take no longer than one day.

Concrete Foundation Walls: Concrete foundation walls are generally specified for houses built on sharply sloping lots or with full basements. The concrete is placed inside wooden or metal forms that must be erected on top or become part of a continuous footing. In addition, more advanced concrete placement techniques, such as vibration, will be required. It is critical, therefore, that your foundation subcontractor be skilled in form erection and placement of concrete within the forms.

The configuration of concrete forms will generally follow the plan view of the house (also referred to as the "footprint") as shown on the site plan. Since the direction and degree of slope of the lot may allow one or more sections of the house to be built at grade, the foundation walls that are required to be formed may not follow the exterior walls. Foundation walls will require a stiff, dense concrete mix. Do not allow additional water to be added on site.

Depending on the complexity of the exterior wall configuration, form erection can take from one to several days. Concrete placement may also take from one to several days.

Foundation Waterproofing: If your house has a basement or other useable space below the first floor, the exterior side of the foundation walls must be waterproofed. Many foundation subcontractors will also install the waterproofing material.

Piling: For building sites that are within a flood zone or have soil conditions that are loose, unstable or of poor composition, a foundation system consisting of piling may be required. Piling subcontractors are highly specialized and, in some areas, very difficult to locate. If your house is to be constructed on a foundation of piling, locate and schedule a piling subcontractor well in advance of the day you will need it.

Piling should be installed under continuous supervision of an architect, a structural engineer or a job superintendent highly experienced in pile location and driving. The local building official may inspect the pilings or may require you to hire a special inspector who is certified to perform piling inspections. Installing piling may take several days or weeks, depending on specific site and soil conditions.

Foundation Drains: As soon as the concrete for your foundation reaches its initial set (usually within 24 hours after the last pour) you can install your foundation drainage system. If your foundation subcontractor will also install the foundation drainage system, no scheduling by you is required. If you must use a separate drainage subcontractor, call it at the same time you call your foundation subcontractor and schedule installation for the day after the footings are poured.

Floor Preparation: If your first floor is to be concrete, the next step in your order of progression is to fill the area with clean soil, compacted to at least the density of the natural soil. Make sure that the compaction procedure and the percentage of density meets the requirements of your local building codes. Common laborers can perform the filling and compacting of the soil. Testing for compaction, however, must be performed by a soil testing service or engineer.

Concrete subcontractors will quite often compact the fill material, but the hourly rate will be much higher than that of common laborers.

Filling and compacting, if done by mechanical means, should take no longer than two days. If your local building official requires a compaction test, order it to be performed the day the compaction is completed.

As soon as the fill is compacted to a satisfactory density, call your plumbing subcontractor. Installation of all interior water service and drain piping must be completed before concrete is placed for the floor system. In addition, all below floor components of other subtrades (conduit for electrical and other wiring, gas piping, ductwork, vent piping, etc.) must be installed at this stage.

A rough plumbing inspection will be required before your water service and drain piping may be covered with fill material. Make sure all work is completed and all pressure tests are on the piping before calling for the inspection. The plumbing subcontractor should be able to perform the rough-in installation and have the work inspected in two days. If ductwork, gas piping or conduit is installed within the fill area, call your local building official and inquire about any inspections that may be required. Upon completion of all work and inspections, your pest control subcontractor should treat the entire area below the floor slab.

As soon as all inspections are completed and approved, you will be ready to install the floor. The concrete subcontractor should be called at the time the plumbing subcontractor starts it's work. Under most conditions, the concrete subcontractor will need several days notice to schedule the work. Make sure the fill material is in place and firmly compacted around all pipes, ducts and conduit, before the vapor barrier and reinforcing fabric is installed.

If a slab inspection is required in your area, call for it before you order the concrete. Make every effort to have the inspection performed on the afternoon before you are ready to place the concrete. By doing so, you will have the time to correct any deficiencies that may be discovered by the building inspector. If the slab inspection is approved, your concrete subcontractor will be able to start work as early as necessary the following day. Unless the first floor is unusually large or complex, the placing and finishing of the concrete should take one day.

At this point, you are ready to call in your framing subcontractor to start the structural framework. If your first floor is to be framed out of wood, your framing subcontractor should be prepared to start the day after the masonry subcontractor completes the installation of the curtain walls and/or piers.

Frame Work: The framing subcontractor should start framing the balance of the superstructure (walls and roof) as soon as the floor system is completed.

Elevator: If your house will contain an elevator, the elevator subcontractor should be called to coordinate its shaft work and car installation with the framing subcontractor.

Plumbing: The plumbing subcontractor should continue (for concrete floors) or start (for wood frame floors) the installation of the water piping and the drain waste and vent piping when the superstructure framing is completed.

Note! It is not necessary for the exterior wall and roof sheathing to be in place to allow the plumbing subcontractor to start work. If your plumbing system is designed with vents that penetrate the roof, the roof sheathing must be in place before the vents can be completed.

Gas: The gas subcontractor should continue (for concrete floors) or start (for wood frame floors) the installation of any gas service piping at the same time as the plumbing subcontractor.

Note! Many plumbing subcontractors also install gas piping. While you are in the process of obtaining bids for the plumbing work, ask the plumbing subcontractors if they also install gas piping. If so, you can usually negotiate a lower price by allowing one subcontractor to perform two tasks.

Wiring: The electrical, telephone, cable television, entertainment and alarm systems subcontractors may start work as soon as the wall and roof framing is in place and should be called in at the same stage of construction as the plumbing subcontractor. Installation of all device boxes, rough-in frames (for intercom and alarm systems), panels and the various types of wire may be performed simultaneously with the installation of the plumbing piping. Wiring for a doorbell if used must be installed at this time.

Mechanical: The mechanical subcontractor should start the installation of the ductwork as soon as the roof sheathing is installed.

Fireplace: If your house design contains one or more fireplaces, your masonry subcontractor should be called in at this time.

Roofing: The roofing subcontractor may start work after the roof sheathing, plumbing vents, chimneys and any other components penetrating the roof (power attic vents, etc.) are installed.

Inspections: Upon completion of the roofing subcontractor's work and with the framing, plumbing, wiring, mechanical, fireplaces and chimneys in place, you are ready for your rough-in inspections. Required inspections may include:

Framing inspection, which will consist of verifying the size, type and grade of all lumber used, as well as spacing, cutting, notching and bracing of all structural members and anchoring of the floors, walls and roof. In addition, all windows and exterior doors, as well as interior stairs and landings will also be inspected.

Note! Some building officials require a portion or all of the insulation to be in place before the framing inspection may be requested. It will save you and the building official time if you determine the requirement for insulation inspection in advance.

Plumbing inspection, which will consist of verifying the size and material of the piping used, the type fittings used, the material used to bond the pipe and fittings together, proper fall (slope) of the piping and general plumbing techniques. In addition, one or more of the following tests may be performed on the system.

Water test, which will consist of plugging all openings in the system, filling the entire drain, waste and vent piping with water and inspecting each fitting connection for water leaks. The water must be contained by the piping for a minimum time period (usually 15 minutes) before the test may begin.

Smoke test, which will consist of plugging all openings in the system, pumping a thick black smoke into the piping with a smoke machine and inspecting each fitting connection for smoke leaks.

Peppermint test, which will consist of plugging all openings in the system, pouring a solution of oil of peppermint and boiling water into the highest vent opening and tightly capping the vent. If a fitting connection emits a peppermint aroma, it is not sufficiently sealed.

Note! The smoke and peppermint tests are now all but obsolete. I have included them only because in some very rare instances they may still be used.

Hydrostatic test, which will consist of capping and sealing all potable water piping, filling the system with water, introducing air pressure (usually 25 pounds per square inch above street pressure) and inspecting each fitting connection for water leaks.

Electrical inspection, which will consist of verifying the size and type of wire used, grounding requirements, location of device boxes, number of outlets per circuit, distance between receptacles, securing of wires and location of the panel(s).

Mechanical inspection, which will consist of verification of duct size, duct material, duct connections and materials and method of connections, location of the source of supply for the return air, location of all registers and balancing of the system.

Note! Combination inspectors (persons trained to make inspections in a variety of trades) normally perform all rough-in inspections during one visit. In some areas, however, inspectors who are qualified and specialize in only one trade may be employed.

Insulation: After you have obtained approval of all systems that are required to be inspected, you will be ready to install the insulation. As stated in Chapter Five, insulation may be installed by the framing subcontractor or by an insulation subcontractor.

Note! Batt insulation may be installed between the ceiling joists at this time. If mineral wool or any type of mechanically blown or loose insulation is used, the ceiling will have to be in place first.

Exterior Wall Finish: If the exterior wall finish of your house is of brick or other masonry, installation should be started as soon after the framing inspection as possible. If wood or other nonmasonry materials are used, you may allow much less time for installation.

Exterior Trim: If the exterior wall finish is going to be masonry, your masonry subcontractor may prefer that brick molding be installed around all windows and doors before the masonry is started. If so, the framing subcontractor should install the brick molding before the masonry subcontractor starts.

Interior Walls and Ceilings: At this stage, your walls and ceilings are ready to be closed in. Contact your dry wall subcontractor and schedule its starting date as soon as installation of the insulation is complete.

Interior Wall Paint: As soon as the dry wall is complete, the paint subcontractor may start painting the interior walls.

Electric: It will be necessary to call the electrical subcontractor as soon as the interior wall paint is complete. The electrical subcontractor should continue its work until the installation of all of the electrical devices has been completed.

Note! Installation of the light fixtures, switch and receptacle cover plates and panel cover will take place after the paint subcontractor is finished.

Garage Door: Garage door installation may be performed at any time from this point forward. If a ceiling is specified in your garage, it should be in place before the garage door is installed.

Interior Trim: The framing subcontractor should now begin to install the interior window and door trim, baseboard, crown, corner and other specialty moldings.

Exterior Trim: As soon as the interior trim is complete, the framing subcontractor should move to the outside and complete any remaining exterior trim work.

Exterior Paint: The paint subcontractor should prime then paint the exterior trim as soon as it is installed.

Note! I prefer to prime and paint the exterior trim material, front and rear sides, before it is installed. Doing so is less time consuming and provides much better protection for the wood. In addition, you can usually negotiate a lower price for painting the trim work if the paint subcontractor does not have to work from ladders or scaffolds.

Mechanical: The mechanical subcontractor should be called in at this stage to complete the installation of the heating and air conditioning system(s).

Floor Covering: Vinyl, tile or wood floor covering in the kitchen, bathrooms (including tiled tubs, showers, etc.) and all rooms not receiving carpet should be installed as soon as the electrical and mechanical subcontractors have finished.

Cabinets: As soon as the floor coverings are in place in the kitchen and bathrooms, the cabinets should be installed.

Note! Some cabinet subcontractors prefer to install the cabinets before the floors are finished. To eliminate problems, make sure you understand at which point your cabinet subcontractor would prefer to perform it's installation.

Plumbing: The plumbing subcontractor can be called in to complete the plumbing as soon as all kitchen and bathroom cabinets have been set. The plumbing subcontractor should continue until all fixtures, including the water heater, are set and properly working.

Wall Coverings: Wallpaper, wall tile and glass (for large or fixed mirrors) subcontractors should be ready to start as soon as the kitchen and bathroom cabinets are set in place.

Electrical: As soon as all the walls are covered, call the electrical subcontractor in to install the switch and receptacle cover plates, light fixtures and panel cover(s).

Wiring: The telephone, cable television entertainment and alarm systems subcontractors, if used, should be scheduled at this point to complete their respective trim-outs.

Elevator Trim: If an elevator was installed, the trim and finished components should be installed at this point.

Carpet: The last step in your order of progression for the construction phase is the installation of the carpet. Once the carpet subcontractor completes the carpet installation, keep all nonessential people out of the house until you are ready for occupancy.

Construction Cleanup: When all of the subcontractors have completed their tasks, you will have the option of hiring a service that specializes in "construction cleanup" or you can perform the tasks yourself. Construction cleanup involves a general cleaning of the house to prepare it for occupancy. Some of the tasks include scraping excess paint from windows, washing excess dry wall mud from moldings, vacuuming floors, wiping all hardware, counter tops, cabinets, etc.

Final Inspections: The structure and all systems within it must receive final inspections from the building inspection department before occupancy should be attempted.

Electric, gas and water services are not normally connected until the respective inspector tests and certifies each individual system. The inspector, authorizing one or more of the services to be connected and turned on by the utility company, may issue tags, stickers or other certificates. In addition, the building official in most areas will issue a Certificate of Occupancy (also known as a CO) before services and/ or occupancy is permitted.

Final inspections usually consist of a visual check of all systems and components such as electric and plumbing fixtures, mechanical equipment, exposed piping, (traps and connections under sinks, etc.) doors, windows and all finished items.

Along with the visual inspections, many inspectors actually test various components or systems. Smoke and peppermint tests may be performed for finished plumbing systems.

Electrical equipment, fixtures, switches and receptacles can be checked by "jumping" current from your temporary service to the house side of your electrical meter base.

CAUTION! Jumping current may not be permitted or even legal in your specific area. Make sure that the electrical inspector and the power company is contacted and permission (preferably written) obtained from both before considering such action. After permission is granted, only a qualified master electrician must make the connection. The current should remain on the system for only the length of time necessary to verify the components or for the duration of the inspection. Do not attempt to use a jumped service for any other reason than to test the circuits.

If the house is without electrical current at the time of the final inspection, the inspector can nonetheless verify the wiring by using one of several meters. One such meter, known as a megger, when cranked, can produce as much as 500 volts of direct (DC) current. When applied to the electrical system, it will indicate shorted circuits, wiring insulation flaws and other system malfunctions. A volt/ohm meter may also be used to detect shorts and further check the system.

Although the final inspections are relatively simple, do not underestimate their importance. Do not attempt to move furnishings or household goods into the house until your certificate of occupancy is in hand or unless you have written permission from your building official.

Determine the requirements for final inspections, service connections and certificate of occupancy about one week in advance of anticipated completion.

At this point the construction should be complete and your house ready for occupancy. Landscaping, patios, sidewalks and driveways, etc. may now be installed or, if necessary, can be delayed for a reasonable amount of time.

RECORDS AND PURCHASING—Complete and accurate record keeping is a must for precise budget and time control. A standard bookkeeping ledger or notebook may be used to keep accurate daily and monthly purchase records.

If at all possible, purchase all of your construction materials, components and equipment yourself. The task requires some time and effort on your part, but assures that you will have control over your material expenditures. Check with your subcontractors on a daily (or nightly if necessary) basis and request a list of the items and quantities that will be needed. Attempt to stay about two days ahead of the subcontractors with your materials ordering. If you elect to authorize any other person or persons (the fewer the better) to purchase materials on your accounts, however, make sure you obtain receipts from them on a daily basis.

Note! Insist that all prices you are charged for materials are shown on your receipts. Some suppliers will only list the materials and quantities, claiming that it takes too much time to look up the price of each item. The supplier must list the prices at your request. Request them! In addition, compare the prices on your receipts with the quotes given to you. If a discrepancy exists, get it resolved before signing the receipt.

Be aware of the materials being purchased for your project. By having knowledge of the type and amount of materials necessary and the daily cost of each, you reduce the possibility of overpurchase and overpayment by a considerable amount.

To keep tract of your expenses, develop a chart to log each and every purchase you make. You can create a simple, yet precise method of logging your expenditures by using a ruled pad or ledger. At the top of each page, create four headings; Date, Supplier, Item(s) and Amount. Every time you purchase an item or component, document the date of purchase, the supplier from which the purchase was made, the item or items purchased and the total amount paid. At the bottom of each page, include a space for the total for the month. To obtain an accurate final cost of construction, it is critical that you log every purchase regardless of size or amount.

In addition to your ledger, obtain two small boxes (shoeboxes will work just fine) to hold your receipts. On the top of one box print the words "Bills to be paid." On the top of the other box print the words "bills paid." Enter the information for each daily purchase, weather paid for or charged, in your ledger. If your items are paid for at the time of purchase, place the receipt in the bills paid box. If your items are charged at the time of purchase, place the receipt in the bills to be paid box. When the invoices for charged items arrive at the end of the month, you have all your receipts handy for comparison. Compare each monthly invoice with your daily ledger entries and the receipts in the bills to be paid box. Pay particular attention to the material quantities and charges on the monthly statement. As hard as it is to believe in this day and age, computers do make mistakes. If you detect a discrepancy on your invoice, take it to the supplier personally. Once the problem is resolved or if no discrepancy exists, pay the bill and place the paid invoice and all corresponding daily receipts in the bills paid box.

Labor paid by the hour can be recorded in a standard pocket sized payroll book. Payroll books have preprinted pages containing all the information necessary for accurate record keeping. All you have to do is fill in the spaces with the appropriate data. Group all workers by their respective subtrades. Payroll books may be purchased in any stationary, office or business supply store. You may also find them in bookstores, stationary sections of department stores and in some building supply stores.

One of the best methods of keeping accurate records of expenditures is by establishing a construction checking account. Establishing a construction account can be accomplished at the bank or savings and loan where you do business.

A construction account will allow you to know the activity of your building funds at all times. You also have the flexibility of adding to the account during construction and keeping it active after completion, for future projects or maintenance.

Setting Up Accounts: Before you start construction, visit all of the building supply stores in your area. Speak with the credit manager or the person responsible for establishing new accounts. Advise that person of your plans to construct a new house and request a contractor's account. Determine the amount of discount that you are entitled to and ask about the policy of returning unused materials. You should get the same discount that professional building contractors are allowed.

Once your account has been established, get to know the store and the salespersons with which you will be doing business. Most stores have a contractor's sales area and salespersons that serve contractors only.

Salespersons are usually paid commissions on total sales. Quite often, bonuses or other incentives are given to the top salespersons in a given period of time, for volume of sales obtained or for meeting specific quotas. Such a system is beneficial to you also, in as much as, a salesperson close to his or her quota may very well reduce prices lower than the standard discount in an effort to gain sales volume. Since the standard contractor's discount is a policy of the store, the percentage of any additional reduction you may receive is deducted from the salespersons commission. The salesperson may be willing to absorb a loss, however, if sales volume is maintained or the bonus is substantial and within reach. Since this bargaining range already exists, do not be afraid to ask for a lower price if the material cost quoted is higher than you think you should be paying. If you happen to receive a reduction of this nature, you should consider purchasing more of your required materials at the same time. Bear in mind, however, that unless your job site is relatively secure or you have a protected area in which to store your goods, you may be inviting pilferage.

Note! A good rule of thumb is to keep no more than three working days worth of supplies and materials on an unsecured site.

Although you can purchase your materials from any salesperson you desire, it will be worth it to you in better service if you choose one person and one alternate at each supply store. Each time you purchase, do so from the same person. In the absence of your regular salesperson, use the alternate. The salesperson and alternate will quickly learn your name and face and in return for sales volume, will give you quicker and better service. Using one salesperson is also an advantage when you have to call in an order late in the day for early delivery the next day. In most cases, instead of hearing "sorry the truck is already loaded" or "delivery will not be until tomorrow afternoon," the response is, "I'll make every effort to get your order out to you." More often than not you will get your order on time. In addition, you can usually continue using your salesperson to purchase materials for other projects or maintenance after the house is complete.

After you make your selection of salespersons and alternates, list their names and telephone numbers next to the store name on a sheet of paper (see the following graph). Make as many copies as you have telephones plus one additional. Post one copy near each telephone and keep the additional one in your wallet.

TABLE 5

Store/Supplier	Salesperson (1) Alternate (2)	Telephone No.
	1.	
______	2.	
	1.	
______	2.	
	1.	
______	2.	
	1.	
______	2.	
	1.	
______	2.	
	1.	
______	2.	
	1.	
______	2.	
	1.	
______	2.	
	1.	
______	2.	
	1.	
______	2.	
	1.	
______	2.	
	1.	
______	2.	
	1.	
______	2.	
	1.	
______	2.	

During the process of establishing your construction account, ask the stores policy for material delivery and returns. Some stores charge a fee for all items delivered to the site. Other stores charge a delivery fee for all orders under a specific dollar amount. Most stores, however, are grateful enough for your business to deliver the materials free of charge. Those are the stores you should seek out. Quite often, all delivery fees are waived for contractors. Since you will be putting a substantial amount of money into the store's cash register, delivery of your materials should be provided free of charge. If your salesperson cannot authorize free delivery, speak to his or her boss and continue, if necessary, all the way to the store manager. If no results can be achieved at the highest level, voice your dissatisfaction and unless the prices are unusually low at that particular store, patronize it as little as possible. Remember that without deep discounts on the materials, delivery fees may use up the difference in price and may even end up costing you more money.

Be sure you understand the policy for returning unused materials before you make your first purchase. Some stores charge a restocking fee for returned merchandise. The charge may be justifiable if a truck from the store is required to go by your site to retrieve the materials. If you or one of your workers return the goods, however, the fee, in my opinion, is excessive and should not be charged.

In all cases, some amount of credit will be given to you for all unused materials. To receive credit, all items must be in their original unopened packages, crates or wrappers.

Note! Monitor the use of as much material that is purchased for your project as possible. Require your materials to be stored in areas that are not subject to damage from workers, vehicles or equipment. Do not allow the material packaging to be opened unless absolutely necessary, weathered or tampered with. Require your workers to use all of the items in a package before opening another.

If any material is delivered to the job site that is damaged or in an unacceptable or unusable condition, reject it. It is best to reject any item at the time of delivery and have the truck driver note the reason for rejection on your copy and the store's copy of the sales receipt.

Before you will be able to obtain credit on unused or damaged merchandise, you will have to produce your original sales receipt. Make sure you keep all of your sales receipts in the appropriate "bills" box and keep the boxes in a reasonably accessible location. Remember that credit amounts to money savings, take advantage of it whenever possible.

Watch your newspaper for special sales at places that will have materials or components that you will need for your project. You can, for instance, find some very good buys at going out of business sales. In addition, most building materials supply stores have what is known as end of year or inventory sales. The purpose of the sale is to clear as much stock as possible from the store before the beginning of the new tax year. In that situation, the store reduces inventory, pays fewer taxes for the goods on hand and you in turn realize some decent savings. Big savings may also be found on product closeout sales or on discontinued items. If discontinued items can be used on your project, they may prove to be exceptionally good values. Original factory warranties and service will still be in effect and will apply to discontinued items for the same length of time as regular priced goods.

Be aware of as many construction sites as possible within your area. More often than not, general contractors have a variety of materials on site that are the result of over ordering. Wood, block, brick, wire, plumbing pipes and fittings and any other type of basic material that may be in broken packaging or partially used lengths, can be purchased at a fraction of the original cost. I have known general contractors to give usable materials away, only because it would be too expensive or cumbersome to move from one job site to another or to put in storage. If you are lucky enough to find a contractor willing to give or sell you materials at greatly reduced prices, (make sure you know the current store prices of the materials before finalizing any sale) your savings could be substantial. Purchases of this nature are "as is" with no possibility for return or refund of money, therefore, be sure you can use all of the materials and, by all means, check all items very carefully for damage before paying for them. Pay for the materials at the time you are prepared to move them (not before) and be sure to obtain receipts for all monies spent. Once you purchase the items, move them to your job site as soon as possible.

You have the same right to be as selective when purchasing from a contractor as you are when purchasing from a supply store. Choose only first quality or usable materials.

If the contractor gives you materials, it is always under the condition that you move the items at your expense. In some cases, the contractor may include broken items or materials that you don't need or can't use. In that instance, you can consider hauling off the unusable items as payment to the contractor for the goods that you can use. Besides, certain broken items (brick, block, open bags of mortar or cement, etc.) may be usable as solid fill material under concrete slabs, for porches or fireplaces, just to name a few locations. In addition, good material that you cannot use can be sold or traded for items that you can use.

Almost all building supply stores have an area or section of a warehouse that contains equipment or components that were special ordered for other projects, but for one reason or another, not used. Naturally, the store had to purchase the items in advance of receiving them. Those materials, therefore, will represent a loss of money to the store if they are not sold. Since the materials are special order rather than stock, they probably will not be in great demand by other contractors. For that reason, the store will sell the items for whatever cost it can recover. Quite often you can purchase the goods at cost or even less if you bargain enough. Ironically, the special order items are usually more expensive or elaborate than stock components. Windows, door units, cabinets, floor coverings, hardware, paneling, molding and many other items may be found in the store's bargain area or warehouse.

Some materials may be purchased direct from a mill, factory, wholesaler or manufacturer. For possible sources, check the telephone book for your area, as well as any major city within a reasonable distance.

Most cities have one or more salvage stores that may sell or even specialize in building materials. Be very cautious about purchasing any material from such places. Common sense should tell you that the sometimes drastic difference in cost for certain materials has got to be for a reason. Usually that reason is because the goods are of low quality, damaged or obsolete. If the quality level is lower than what the building official is willing to accept, your money will be ill spent. Damaged or obsolete items speak for themselves. In the construction business it is wise to remember that "cheapest" does not always mean the most economical or even the least expensive in the long run. As a rule, I do not recommend shopping in salvage stores for components other than trim, wallpaper, paint or other items used for cosmetic purposes only.

Internet Shopping: Many sources for purchase of construction materials are available on the Internet. Although the Internet may be an attractive and convenient tool for shopping for most consumer goods, I tend to avoid it when purchasing construction materials. Components such as light fixtures, plumbing fixtures, wallpaper and roofing shingles, when you know the specific brands, models, colors, patterns, etc. may be suitable for purchase via the Internet. Materials that should be selected or may potentially be returned, however, such as framing lumber, framing hardware, sheetrock, electrical and plumbing system components, molding and millwork, paint, etc. should be purchased through local material suppliers.

Since Internet shopping does not allow you to visually select your materials (other than by photograph) or assure their quality, any purchase would be risky. You must know and trust the reliability of the sources and be convinced that quantity and quality of the materials or components will be delivered on time. If delivery does not occur at the specified date and time, costly delays or work stoppage may occur. At that point you must drop what you are doing and attempt to locate your goods. With Internet purchases, the process is much more convoluted than with local sales. You cannot simply pick up the phone or jump in your vehicle and be talking with your salesperson in a short amount of time. Another drawback is that payment for Internet sales must be made in advance of delivery. The most common form of payment for goods purchased on the Internet is by credit card. The amount of many routine construction material purchases may be large enough to exceed a consumer credit card limit. Also, with some Internet companies, all sales are final and no credit on damaged materials or components will be issued. For those Internet sources that will take back materials or issue credit on damaged materials or components, the process is also convoluted. Construction accounts cannot be established and more accounting and paperwork will be necessary to keep accurate tract of your purchases. Even under ideal circumstances, you must still know the current costs of the materials at the local level to be able to determine if any savings will be realized.

Shopping for construction materials is no different than shopping for any other consumer item. To purchase an item solely because it has the only name you recognize is foolish and may cost you more money than necessary. Good building supply stores have several samples of the same component produced by different manufacturers. Look at them all. Take the time to check the features, colors and styles available for each item you are shopping for. Only after you are convinced that the item will function properly and within your expectations when permanently installed, should you agree to purchase it. Shop around before you make your commitment to purchase. Make sure the supplier you purchase from is the one that gives you the best buy on the item.

Lower priced materials may be used as alternate components in certain areas of your project, producing a cost savings without jeopardizing structural integrity or appearance. For example, jointed trim boards or moldings that are to be painted, serve just as well and give the exact same appearance as solid sawn or single cut pieces, and; utility grade lumber has the same structural value when used as shoes, plates and purlins, as number two or even number one grade lumber of the same size. Do not, however, attempt to use a material of lower grade or quality than your building official is willing to approve for any specific application. Doing so will result in you having to replace or strengthen the components, costing you additional time and money. At that point, the savings that you could have had, disappear and turn into additional expenses.

Do not be afraid to be innovative. Do not reject the use of a brand new product or material simply because you have not seen or heard of it or you don't know of anyone else who had used it. Check the product thoroughly. Look for product evaluation or test reports issued by nationally recognized product evaluation entities. Ask your local building official for advice or guidance concerning any new product or component that you may be interested in using and verify that it will be accepted if installed.

SAVINGS DURING CONSTRUCTION—Throughout construction, the predominate thought in your mind should be attaining your goal of producing the best quality house within the budget you have established. As you move through the construction process, you will realize that project coordination and purchasing are not the only areas in which savings can occur. The more aware you are of your project, the more ideas on savings you will develop. Start with the basic tips given in this section and build from that base.

For instance, you can save money by getting your whole family involved in the construction project. Evenings and weekends provide ideal times to gather the family at the job site for clean up and material reclamation.

Note! I do not recommend the use of very young children or very old adults for the performance of any task on a construction site. Any person you enlist to assist you on your project site should be mature, responsible and physically well coordinated. The last thing you need during construction is an accident or injury involving a person you asked to assist you. In all cases, before you allow any friend or relative to work on your job site, verify their coverage and protection with your insurance agent.

Using your family members to periodically clean your job site can produce savings in several ways. Consider the fact that it takes at least one and in most instances, several persons to keep an average home construction site clean. Those persons, if on the payroll must be compensated by you. If the people that are assigned to clean up tasks happen to be skilled or even semiskilled workers, the cost per day could be substantial.

For Example: The skilled carpenters working on your project may be employed for $18.00 per hour (a low figure in most metropolitan areas), and, the semiskilled carpenters may be employed for $9.00 per hour. If only one skilled carpenter and one semiskilled carpenter spends one hour at the end of the workday for clean up purposes, you will spend $27.00 for the honor. In addition, two man-hours are lost from actual construction time. Add the labor for clean up to the lost time and that hour cost you $54.00.

Any person who can pick up small pieces of materials or hold a broom can perform the same clean up tasks as the highest paid worker on the project. A clean job site will considerably reduce the possibility of accidents and injury, as well as provide an uncluttered environment for your subcontractors to maneuver and work.

During the clean up process you can also collect and restack materials that will get scattered around the site. If the materials are in their proper places, the subcontractors will use them. Materials such as wood, brick, block and even nails can be collected in the evenings or on weekends and used the very next workday. Materials that are not collected have a tendency to deteriorate or get broken during the construction process. At that point, the materials that should have been installed in your building are not useable or returnable and of no value to anyone. Materials that are paid for, if not used or returned for credit, represent wasted money. In addition, you will have to pay someone to haul it off your property, as trash at the end of the project.

Wood in whole or partial pieces should be stacked in piles according to their individual sizes. Even short pieces can be used for purlins or blocking. All partial pieces of wood, if not usable on the job, should be stacked in an out of the way place and used later as firewood if you have a fireplace or wood burning stove.

Bricks in whole units should be gathered and stacked near the areas where the masonry subcontractor is working. Half units of brick should be piled outside each window and door or in any other area where rowlocks (half bricks laid sideways and edgewise) are to be used. Masonry subcontractors will not slow their own progress down to pick up half bricks for use. Whenever a half brick is needed it is normally broken or cut from a whole unit. Since the masonry subcontractor is paid by the total number of bricks used, it is not likely that the masons will collect and use the half bricks on their own volition. If the half bricks are collected and piled for the mason, however, they will be used, especially if you agree to it in advance.

In addition to reclaiming, you should also stack the materials at various convenient areas around the job site for use by the workers the next workday. When the materials are close at hand, the workers can devote more time to installing and less time carrying. If possible, confer with your subcontractors before they leave the project for the day and determine which materials will be used the next day and the specific locations on site. During your evening or weekend reclamation periods, you can stack the materials at those locations.

Odd and dirty jobs are usually good sources for cost savings. Since virtually nobody enjoys doing the odd or dirty jobs, the cost of labor may increase and/or the time to perform the tasks may lengthen. Find out which tasks cause your subcontractors the most heartburn and, if at all possible, offer to do them yourself. Before you offer to perform a task, be sure that you are willing and capable of completing it and that your subcontractor agrees to a cost reduction on the contract price or on a per hour basis in return for your labor. You will soon find that although the tasks may be dirty, tough or both, the savings you realize will be most gratifying. Any time you agree to a cost reduction, get the agreement in writing.

To save money, offer more money! Sounds unusual doesn't it? Quite often contractors will offer their subcontractors a bonus for completing a certain phase or all of it's contract before the agreed upon completion date. The old adage "Time is money", can certainly be applied to every construction project.

A bonus can be a definite incentive for a subcontractor to complete its task much quicker than the time specified in the contract. Properly applied, a bonus will net the subcontractor more money for completing the same job in a shorter period, while saving you money in time and interest on your construction loan.

Be creative when offering a bonus. Bonuses do not have to be negotiated in terms of dollars. Anything of value to the subcontractor may be used as a bargaining medium. You may consider bartering goods, labor or services that you can provide to the subcontractor in exchange for performance of all or part of its contract. Be sure that you offer a value that is fare to you and attractive to the subcontractor. If you are in business for yourself, (especially professional or service oriented) you may pick up a new client as a side benefit.

Before considering any bonus, calculate the value to you in dollars saved, compared to the additional amount that you will have to pay or preferably barter. Offer the subcontractor a percentage of the money saved, but make sure that you are the one that benefits the most from any bonus agreement.

A final note about savings during construction is to secure all tools and materials that may remain on the job site after working hours. Even in the best of neighborhoods materials and tools will be stolen from construction sites. It is odd how people who would never consider stealing a thing under normal circumstances will help themselves to building materials and tools from active building sites. Use whatever method or means available to you to protect your investment and the investments of your subcontractors. Although you will have insurance coverage for theft, a deductible will apply and the initial cash outlay for the stolen items may not be recovered for some time.

PUBLIC OFFICIALS—Throughout this book, reference has been made to a variety of public officials. The reason for the many references is because of the involvement of local government in the building construction industry.

Although the regulations, building codes and construction standards seem complex and the offices and individuals you have to visit bureaucratic, it all exists to benefit you and all future occupants of your building. Each person you come in contact with is there to assist, protect or provide a service for you. It should give you a strong sense of security to know that each inspector that visits your job site is there to safeguard your well-being.

Building inspectors, through the authority of their positions, have a tremendous responsibility. You should at no time feel intimidated by them. As a matter of fact, they will be the first ones to advise you that they are there to help not hinder your efforts. By the same token, do not at any time attempt to hinder their efforts. If you have any questions or problems with any phase of your project, consult with the proper inspector for guidance.

Cities, townships and counties normally employ at least one, but usually several inspectors to inspect work performed within the building, plumbing, electrical, mechanical and gas trades. All inspectors who specialize in a specific discipline fall under the direct supervision of a single person known as a chief inspector. The supervisor for all of the chief inspectors is commonly known as the building official. In some jurisdictions, however, the building official may have a different title established by the local government. In small or rural jurisdictions, the building official may also perform inspections in one, several or all of the disciplines.

In most jurisdictions, the building official is usually a professional person such as an architect, engineer or licensed contractor. In addition, various national entities certify building officials and inspectors by examination. Many cities and counties and some states now require building officials to be certified as a prerequisite for employment. In addition, some states are even requiring building officials and their inspectors to be professionally licensed before they may be considered for employment.

The principal responsibility of a building official is to assure that buildings are constructed to minimum, yet safe, standards. That person is in position to assist you, therefore, you should feel free to consult with or seek his/her advice whenever the need arises.

Every effort should be made to have yourself or the subcontractor, whose work is being inspected, on the job site when the various inspectors perform their inspections and tests. If you cannot be there or have representation, however, the inspector will most likely leave a report of the inspection in the form of a tag, ticket or stamp.

Tags and tickets are the most common and, for easy identification, are usually colored green to designate approval and red to designate disapproval. Rubber stamps indicating the result of the inspection may also be used. If so, the stamp is normally applied to a designated section of the building permit placard or on a separate inspection record card. Ask your building official which method is used in your jurisdiction and become familiar with the process.

If at any time during the course of your project a red tag or rejection stamp happens to appear, don't panic. A red tag or rejection stamp is merely a notice that a code discrepancy exists in a specific system, component or area of your project or that a policy or procedure has not been properly followed. In either case, contact the inspector that performed the inspection (the inspector's name and/or signature usually appears on the bottom of the tag or stamp) and thoroughly review the reason for rejection, making sure that you understand the violation itself and the trade that is involved. Carefully follow all instructions that are provided on or with your inspection tag or given to you by the inspector. If the instructions are not clear or you do not understand any portion of them, consult with the inspector before attempting to continue.

Require the subcontractor involved to correct all discrepancies noted on the tag. Once complete, inspect the corrections yourself. If after close scrutiny you believe that the violations have been rectified, call the inspector that originally rejected the work and request a reinspection. Make every effort to be at the job site along with the subcontractor whose trade was involved when the inspector performs the reinspection.

A green tag indicates that the work for which the inspection was called has been approved. Only after an approval is given for the prior work, may the task be continued or completed.

Refer to the checklist of subcontractors you developed in conjunction with chapter five and prepare an order of progression for your project based on the trade and task sequence for each subcontractor.

Develop a checklist to keep your project in sequential order. To use the checklist, simply check or enter a date in a blank space after each item upon its completion. The following text contains all the elements necessary to develop a comprehensive Construction Checklist. For simplicity and ease of use, you may prefer to eliminate the elements that do not apply to your specific project. The words that are shown in Italics correspond with the information that you may include on your checklist.

Your checklist should include all elements required to accomplish the actual construction portion of your project such as, *construction checking account established, store accounts established, salespersons selected, records ledger established, payroll book established, clearing, preliminary grading and site work completed, excavation completed, foundation inspection approved, foundation installed, foundation walls/ piers installed, rough plumbing installed, plumbing inspection approved, gas piping installed, gas inspection approved, electrical and other conduit* (below slab) *installed, electrical inspection approved, mechanical duct installed, mechanical inspection approved, fill and tamping for basement or first floor slab completed, soil treatment completed, vapor barrier installed, concrete slab reinforcing material installed, concrete slab inspection approved, framing completed, fireplace/chimney installed, telephone installed, roof covering installed, plumbing drain, waste and vent piping installed, electrical wiring installed, other wiring installed, mechanical duct* (throughout the house) *installed, gas piping installed, framing inspection approved, plumbing inspection approved, gas inspection approved, electrical inspection approved, mechanical inspection approved, insulation installed, insulation inspection* (if required) *approved, garage door installed, drywall/plaster/paneling installed, ceramic wall tile installed, interior trim, completed, exterior wall covering completed, vinyl/ceramic tile/brick/wood floor covering installed, cabinets installed, plumbing fixtures installed, gas appliances installed, mechanical equipment installed, electrical devices and light fixtures installed, interior paint completed, exterior paint completed, wallpaper installed, alarm systems installed, intercom system installed, vacuum system installed, other special systems installed, carpet installed, fine grading/site work completed, landscaping completed,* and *driveway/sidewalks/patios/decks completed.*

CHAPTER TEN

PROJECT COMPLETION

The most exciting part of a construction project is its completion. To view the finished product and know that in a very short time you will be able to enjoy the fruits of your labor is a thrill that is very difficult to duplicate. At this point, every bit of the project is visually and functionally brought into prospective.

FINAL INSPECTIONS—Most owner/contractors tend to be confused as to what constitutes the completion of the project. Building officials usually define complete, as meaning the building is ready to safely occupy. The final inspections, therefore, are the last items in your order of progression and the very last function performed before you occupy the structure.

Discuss the requirements and procedure for the final inspections with your building official approximately three weeks before you anticipate the completion of the project. Doing so will insure that you understand the procedure and level of completion necessary for occupancy. Issuances of many certificates of occupancy have been delayed because of poor timing or misunderstanding of the final inspection process.

Depending on the requirements of the jurisdiction in which you are building, your final inspection may range from a simple walk through the house by a single inspector, looking for obvious code violations or uncompleted work, to a very complex series of tests of the various systems by several inspectors. A comprehensive inspection may include actual operation of all built-in appliances and the heating and cooling equipment; testing of all electrical devices and fixtures; testing of the entire plumbing system and fixtures and a review of all records and prior inspections.

PARTIAL CERTIFICATE OF OCCUPANCY—On occasion, building officials will allow occupancy of a residence before it is totally complete. In such instances, the homeowner is issued a partial or temporary certificate of occupancy.

A partial certificate of occupancy allows the homeowner the flexibility of completing some of the work after the home is occupied. If issued, a partial certificate of occupancy may contain various restrictions or pertain to use of specific areas of the building only, with the remaining areas physically blocked or barricaded.

Partial certificates of occupancy are useful instruments if materials, components or fixtures needed for completion, but not absolutely necessary for occupancy, have not been delivered to the job site. The building official will not issue a partial certificate of occupancy, however, unless the house is safe to occupy. The building official, as well as the homeowner must employ a considerable amount of common sense, before occupancy of this nature is permitted.

CERTIFICATE OF OCCUPANCY—The actual or permanent certificate of occupancy will be issued when the project is completed. It is an extremely important document and should be stored in a safe and secure place.

Your bank or lending organization, insurance carrier and utility company may require the certificate of occupancy, to demonstrate compliance with the required building codes and proof of project completion.

In some jurisdictions a certificate of occupancy is issued only upon request of the contractor or property owner. If this is the case in your community, request the certificate. It is and will always remain the only proof that the house passed all of its required inspections and was certified as complete.

In most instances, an application must be prepared before a certificate of occupancy can be issued. If an application is required in your jurisdiction, obtain, complete and submit the form at the time you call for your final inspections.

UTILITIES CONVERSION—Conversion of your electric and water utilities, from temporary to permanent service, can take place at any time after your final inspections. The electric meter base is usually stamped, tagged or marked after approval by the electrical inspector.

Natural gas and sewer service, if required, should be applied for at this time. Make application at the business office of the appropriate utility suppliers serving your area.

All deposits paid for temporary services, should be carried over and applied to the permanent services at the time of conversion.

INSURANCE CONVERSION—Before you occupy your new residence, make sure that your insurance is converted from the construction policy to a homeowner's policy.

TEMPORARY TOILET—After all construction tasks have been completed and you are sure that no more workmen will be on the job site, you should cancel the rental of your temporary toilet facilities. You may be tempted to cancel the toilet service when one or more water closets within the residence are connected and placed into service. Resist the temptation! The money you save in rental fees may be spent for additional cleanup or damage that may occur to the finished areas of the house.

SAFETY AND CAUTION—Practice safety and caution throughout the entire construction process as if it were a religion. Doing so, especially during the finishing stages will save you cold, hard cash.

Install the windows and doors as early in the construction process as possible and lock them every night and at all times when you or your workmen are not on the site. In addition, whenever possible, store materials inside the locked building.

Restrict people, especially curiosity seekers and those who are "just looking around" from roaming through the building. Doing so will reduce the chances for injury and theft of tools and materials. As rooms and areas of the house are completed, lock or physically block them from entry by unauthorized persons. Allow entry to only those workmen who must be in the rooms or areas to complete specific tasks. Hold all persons that enter the restricted areas responsible for any damage or soiling that may occur.

MOVING IN—At this point your new house is ready for occupancy. All of the work, time and energy you spent is about to pay off.

You have coordinated construction of the building yourself and saved a considerable amount of money in the process. In addition, you have tackled and successfully completed a project that intimidates even the handiest of the do it yourselfers. You will have an everlasting and very well deserved pride in your accomplishment and the finished product.

Take a deep breath, smile and feel good about yourself, because, now it is time to make your house a home.

Develop a Project Completion Checklist to maintain a sequence of events. To maintain a time schedule for future reference, enter the date in a space provided after each task is completed. The following text contains all the elements necessary to develop a comprehensive Project Completion Checklist. For simplicity and ease of use, you may prefer to eliminate the elements that do not apply to your specific project. The words that are shown in Italics correspond with the information that you may include on your checklist.

Your checklist should contain elements such as *final building inspection approved, final plumbing inspection approved, final gas inspection approved, final electrical inspection approved, final mechanical inspection approved, partial certificate of occupancy issued, certificate of occupancy issued, temporary electric service converted, temporary water service converted, gas service installed, sewer service installed, insurance policy converted, temporary toilet returned* and *the date of occupancy.*

CONSTRUCTION PROCESS SUMMARY—In theory, we have been through the construction process for a typical house from start to finish. By now you should have a good, solid understanding of the elements of the construction process and how they relate to each other. But, although your understanding may be solid in theory, you are still inexperienced in its practical application. As in all theoretical examples, the process occurs under ideal conditions and appears to be easy to employ. Actual application of those theories, however, happens in real time, under real conditions and involving real people and real money. Although the construction process is not insurmountable for an inexperienced person, it is far from easy. Our theoretical house was built without consideration given to weather, material shortages, accidents, problems with subcontractors, etcetera, etcetera and etcetera.

The role of the owner/contractor is to coordinate and manage the construction processes for a building for personal use. It is a serious, sobering and often humbling experience. Look at that role, not as a do-it-yourself project, but with the same intensity as if it were your business. In fact, it will be your business. You will manage, control and be accountable for every aspect of the project. You will also be responsible for completing the project on time and within budget.

My motivation and goal in writing this book was to save its readers money. Your motivation and goal in reading it should be the same. In the preface I stated that by the end of this book, your decision to tackle a construction project as an owner/contractor would be clear there will be no maybe. At this point you may have even made your decision. Before you lay the book down and act on your decision, however, bear in mind that you have not reached the end. You have two critical chapters left to read. Read them!

CHAPTER ELEVEN

A CASE HISTORY

It is typical for a book of this nature to have a variety of case histories or testimonials scattered throughout its pages, touting the merits of the theories presented in its text. Case histories offer proof that other people diligently followed the author's advice and succeeded in accomplishing the book's objectives. When I first considered incorporating testimonials into this book, I interviewed several people who successfully used my theories with varying results. After considering my options, however, I decided to use one chapter length case history, as opposed to several multi-page testimonials.

As is the situation with all information or quotes taken from other individuals, accuracy must be verified before inclusion in a book. One advantage to using a single testimonial is that only one source must be verified. Verification of the source and facts in this particular case history, however, was very easy. In fact, I will attest without reservation, that every detail in this case history is true and correct, for it is an account of my own, latest home construction project, taken directly from my handwritten notes and entries in my construction diary.

The purpose of providing my own case history is to illustrate actual issues and conditions encountered on an actual construction project, and explain how they were addressed. This example will take you out of the sanitized, by-the-book realm of theory and place you into the real world of construction management. It will also illustrate the point that even the most experienced of contractors have issues that occur and must be addressed on a daily basis. How smooth and efficient your project runs, therefore, can be determined by how effectively you handle the anomalies as and when they occur.

As you read about my adventure, you will notice that I distinguish various events with the word "Situation" in bold letters. After you read each situation that is presented, stop reading, put yourself in the role of the contractor and think about the issue, then mentally determine how you would have handled the situation. Following each situation, I provide the decision I made. Some situations may be simple to resolve in your mind or even easily overlooked, others will require some thought. In all instances, a decision must be made. Use the information as an exercise to help prepare you for some of the many situations that can and will occur on your construction project.

To add interest and a bit of levity to my case history, I use fictitious company names for some of the subcontractors that I employed. Those names are not the actual names of the subcontractors I used. Furthermore, at the time I constructed my house, those company names did not exist in the greater Columbia, South Carolina area. Any company or business name that is identical to those used in this chapter is purely coincidental.

ANOTHER DREAM HOUSE—Sure, we knew it would happen. It was just a matter of time. Then, one sunny, Saturday morning without fanfare, BAM! Empty nest syndrome hit us like a ton of bricks. That morning, my wife Joan and I came to the conclusion that we lived in a house with much more space than we needed and it was time to downsize. Within a couple hours, we were making our plans to build yet another dream house.

After spending several months looking at the amenities available in and around Columbia, South Carolina, we determined the general area in which we wanted to build our new house. We sold our then existing house, rented another in our area of interest and started our search for property in earnest. During our search, we passed one particular subdivision several times, but did not drive through it. As our area of interest became smaller and more focused, however, we finally drove through that overlooked subdivision and, you guessed it, found and eventually bought our lot. That was on September 20 (the year is not important).

THE CHALLENGE—As far as South Carolina weather is concerned, the perfect time of the year to start a construction project is during the early fall. The blistering heat and stifling humidity of summer is past and the days are still long enough to make good progress. Best of all, the low volume of rainfall allows for minimum downtime. My challenge, therefore, was to design a house, prepare the plans, obtain all approvals and permits and get my project started during that precious time period. Oh, by the way, at the time I was employed on a full time basis by the State of South Carolina as the Administrator for the Building Codes Council, which meant the project had to be managed on my own time. Nonetheless, I was off and running.

PLAN PREPARATION—The lot was one quarter of an acre in area with a rise in grade of approximately six feet from the street to the front side of the buildable area. The rise of the natural grade, although not high, occurred in a short distance resulting in a steep hill at the front yard. My initial task was to develop a site plan to determine how I was going to use the lot. Using a Computer Assisted Design (CAD) program, I prepared a preliminary site plan showing an "L" shaped house with a swimming pool and large patio in the rear, a circular driveway with vehicle parking in the front and various landscaped spaces for the front, rear and sides. To use the steep grade to my advantage, I designed two descending landscaped terraces in the center of the circular driveway. I hate mowing grass, so I used as little as possible in my site design. The site design only took two days to prepare.

The next task was to produce a set of plans and specifications for the house itself. I performed my own plan preparation using the CAD program. The house was a single story traditional design with a two-car garage and a large finished "room over." It had an exterior frame of eight-inch concrete blocks and an exterior finish of brick. The first floor was concrete, the interior walls were wood and the floor for the room over the garage and the roof system utilized wood trusses.

Although I felt comfortable with my design, before I submitted the plans to the Architectural Review Committee for the subdivision, I wanted additional professional input. I contacted two non-associated professional house designers and asked them to review and critique my drawings. In less than one week I received several good comments and suggestions from both. I incorporated most of their suggestions into my plans. Upon completion of the plans, I prepared a disk for printing and had four full size sets produced by an architectural plan printing service. Preparation, revision and printing time for the house plans took approximately six weeks and I was already into November.

Since my design incorporated a masonry frame exterior wall system, I thought it would be wise to start looking for masons that could construct the walls. Most masons that are qualified to construct masonry frame walls, at least in the Columbia area, prefer to work on the larger more lucrative commercial projects. Small residential projects are a nuisance that they would rather not contend with. After many telephone calls, I found two qualified masonry contractors that agreed to bid on my project. I met with the owners of both companies, gave them each a set of plans for bid purposes and waited for my written estimates.

SEEKING PERMISSION—My personal preference when conducting business is to meet with people face to face unless it is physically impossible. I made an appointment to meet with the architect who chaired the Architectural Review Committee for the subdivision on Monday, November 4, at 9:00 AM. When I arrived for the appointment, the chairman's assistant informed me that he was sorry, but the chairman had personal chores to take care of that morning and asked me to leave the plans with him. I left the plans and a check in the amount of four hundred dollars for the plan review fee, with the assistant's assurance that they would be reviewed within a day or two.

Two days went by without a word about my plans, so I placed a telephone call to the chairman. The assistant answered and informed me that the chairman was in a meeting and would be "tied up" for the majority of the day. He took my telephone number, however, and vowed to give it to his boss. No return call. On Friday, I called again to inquire about my plans and again was informed by the assistant that a return call would be forthcoming. Late that same afternoon, the chairman returned my call, apologized for the delay of my plan review and agreed to meet with me at 9:00 AM, at his office the following Monday.

Monday morning, November 11, I was greeted by the assistant and informed the chairman was in a meeting that was running a little longer than originally anticipated. I waited about 30 minutes and finally got to meet with him. During our meeting, the chairman informed me that the plans looked good, but he wanted several minor changes. He also said he needed actual samples of the brick, roof shingles and exterior paint that I planned to use. The balance of the week was used to make the changes to the plans and obtain the brick and shingle samples. I threw the two remaining sets of plans in the recycle bin and had four new revised sets printed.

The following Monday, November 18, I made an early trip to the chairman's office to deliver the plans and samples. The three trips I made for the architectural review process added 189 miles to my truck and, since they were during normal workdays, totaled four and a half hours of annual leave. At that point, It looked as though I would be able to start construction, a little later than I expected, but still in time and take advantage of some of that nice, dry weather. I missed most of the fall season, but that didn't bother me all that much since

South Carolina was in a drought that was extending into its third year. Rain was the least of my concerns. Besides, excitement always runs in high gear just before the start of a construction project. I was ready to go.

LOT PREPARATION—I hired "Tractor's by Danny," to clear and rough grade the lot. Before the grading subcontractor could start, however, I had to do some light clearing to expose the property corners. Danny informed me that he could do all of the required work the following Monday.

THEN IT RAINED—On Friday, the weather took an awful turn for the worse. It rained with a vengeance for several days without letting up. The ground was saturated. Everyone was relieved and happy to finally see a nice, long drenching rain to quench the parched earth. Everyone, that was, except me!

The soil for my building site was clay. What we refer to in South Carolina as "gumbo" clay. When it is dry it is as hard as concrete and when it is wet it is extremely slippery and instantly sticks to everything, including itself. You take one step on it; it will stick to your shoes. You take another step and the clay sticks to itself. Several steps later your feet are as large as concrete blocks and just as heavy. Gumbo comes in colors and stains as quick as it sticks. Mine was mostly an orange-reddish color with random veins of gray.

SEEKING MORE PERMISSION—While it was raining, I decided to obtain my building permit. Before I could make application, however, I had to pay the sewer tap fee to Richland County. I took a half-day of annual leave to pay the tap fee and start the permit process. With the tap fee receipt and building plans in hand, I went to the Richland County Building Inspection Department, to get a building permit application.

Several years had gone by since I built my last house in Richland County. I was aware that changes in personnel and many of the administrative processes had occurred since my last project. I was not aware, however, of the magnitude of those changes, but I was due to find out.

Although I held licenses as a commercial contractor and residential homebuilder, I applied for my building permit as an owner/contractor. The permit clerk informed me that, since I wanted to list myself as an owner/contractor, I would have to provide several more details on my building plans. Knowing that I already prepared a very detailed set of building plans, my curiosity got the best of me and I had to ask the inevitable question. Why? I was informed that owner/contractors in general, were not familiar with the building codes and construction processes. Consequently, they would start a project with insufficient plans or construction knowledge, then call the building inspectors asking for design or construction advice. In an effort to give the inspectors maximum time for their inspections, it was the department policy, therefore, to require more detailed plans for people building as owner/contractors. At that point I mentioned that I held two state issued construction licenses and was fully aware of the building codes and the various construction processes. The clerk informed me that if I desired to apply as a licensed contractor, the plan approval process would be waived. I provided my license number, got my building permit application and headed home.

I constantly stress the point of not paying money without having goods in hand or a service performed. I broke that tradition on Wednesday, December 15, when I contracted with Diamond Pools of Columbia and placed a down payment for a swimming pool with its sales representative Pat. That action confirmed my theory and proved to be the single biggest mistake (and regret) for the project. Meanwhile, both of my prospective masonry subcontractors finally called me. Their messages were identical. There were not enough skilled masons in the greater Columbia area to keep up with the volume of work that was occurring at the time. One company informed me that it could not take any more work and the other submitted a bid so high, that it removed itself from consideration. I headed back to the computer and redesigned the exterior walls, replacing the masonry with conventional wood frame.

LOT CLEARING—Rain delays cause a backup of work for all of the subcontractors that perform any part of their trades outside. When backups occur, the subcontractors will reschedule their projects, with the contractors that give them the most work getting top priority. Although I was performing as a licensed contractor, my subcontractors did not consider me a source of continued work. When the ground dried enough to start the clearing, I got shoved to the end of the priority line. On Thursday, December 19, Danny, my grading subcontractor finally got around to clearing my lot. Then it rained some more.

During the clearing process, the blade on Danny's bulldozer cut a little too deep and caught the water service line running from the street to the lot. Water was shooting out of the main line like Old Faithful, with no way to stop it. Finally, Danny was able to crimp the pipe and tightly wrap the crimped end with bailing wire. The next day I reported the break to the water department and asked if they could put a rush order on installing my water meter. My water meter was installed several days later. I attached a water spigot to my water meter and placed it in an area where I was sure it would be away from the actual construction, but readily accessible to the workmen.

With the trees gone and the lot fully exposed to the rain, my gumbo turned to orange reddish slop. I needed a good stretch of dry, breezy weather to evaporate the water and solidify the soil before I could go any farther. While I was waiting, I decided to follow up with my prospective masonry subcontractors and complete the building permit process. When I contacted the masonry subcontractors, both informed me that they were too busy to do a takeoff on my plans. I was assured that the estimators would work my plans in within a week.

STILL SEEKING PERMISSION—I was into the Christmas week and many businesses were down to skeleton crews. On Monday

December 23, I was able to go to the building inspection department and walk right up to the counter with no lines to contend with. I handed my completed building permit application to the clerk and was prepared to write a check for the fee. Instead, the clerk informed me that before the building permit could be issued, I needed to provide a list of the major (electrical, plumbing and mechanical) subcontractors that would be working on my project. In addition, since I was to be listed as a general contractor on the permit, I needed a building permit bond.

The purpose of a building permit bond is to provide the project owner with a source of money, if it could be proven to the bonding company that the general contractor was overpaid for the value of the work performed, and then absconded before the project was completed. It was obvious, at least to me, that if I was the builder of record, building a house for myself as the property owner, I certainly was not going to cheat myself out of money, run away and never complete the project. I asked the clerk why I needed a bond to protect myself from myself. The clerk informed me that it was the policy of the building inspection department to require all general contractors to provide permit bonds as (limited) protection for their clients, no exceptions. She then informed me that I could apply as an owner/contractor if I desired. After several minutes of meaningless dialog about how illogical a building permit bond would be in my case, I realized I was only wasting time. I took my building permit application, along with my new instructions and left the building. As I walked to my truck, I wondered what happened to common sense.

Obtaining a building permit bond is quick and easy. I called a bonding company, completed the paperwork, paid the fee and had my bond in my hands on Monday, December 30. I felt much better and certainly more secure, knowing that I had some monetary protection from my potentially unscrupulous contractor.

As soon as I make a commitment to build, I focus on saving money during every phase of the process. Material sales occur all year and I constantly look for them. During my trip to the building inspection department, I happened to see a sign advertising an end of year sale on granite countertops. Tuesday, Joan and I went to the granite sale, found exactly what we wanted, negotiated and purchased all of our kitchen counter tops from Chad, the co-owner of "Naturally Stoned" tile and countertop supplier. I made my first major purchase for a "finished" material before I broke ground.

I had already performed enough research to know which major subcontractors I was going to ask for bids. Not many subcontractors work the week between Christmas and New Years day, so my initial contact with them had to wait until after the first of the year. During the next week, I set up 12 appointments and met with the subcontractors individually on the project site. I was attempting to save as much of my annual leave time as possible, so I set the appointments during my lunch hour, after work and on weekends.

On Friday, January 10, the sun was shining and the temperature was balmy. Joan and I closed on our construction loan and I was bursting with optimism. The rain continued, on and off, but mostly on, however, for most of the month of January.

GETTING PERMISSION—The big day finally arrived. I had my list of subcontractors, my building permit bond, my building permit application and a partly cloudy day. It was Friday, January 24, and I was ready to obtain my building permit. The permit clerk was in a cheery mood and everything was going well, until I saw a small scowl appear on her face. She informed me that the business license for the electrical subcontractor expired and had not yet been renewed. I was determined to obtain my building permit that day, however, so I paid the electrical subcontractor's business license fee. The building permit was issued.

LAYING THE FOUNDATION—The soggy earth needed several days to dry before it would support the weight of a backhoe. I hired "Bill and Bill," a father and son team that billed themselves as foundation subcontractors who, as it turned out, were more qualified as bull artists than anything else. Bill and Bill set the batter boards and dug the footings on Friday, January 31. Later that afternoon it started raining again. The footing trenches filled with muddy water, but the foundation subcontractor wanted to pour the next day anyway.

Situation—Concrete diluted with rain or trench water would cure to a lower compressive strength, resulting in a weaker foundation. The foundation subcontractor wanted to complete the pour or move to another job and reschedule mine for a later date. More rain was forecasted for the following week. If you were the general contractor, what would you have done?

My Decision—I cut relief drains into the footings to drain the water, increased the strength of the concrete from 3,500 to 4,500 PSI and required a stiffer (dryer) mix. The combination of increasing the compressive strength of the concrete and specifying a dryer mix compensated for any loss of strength that may have occurred due to dilution.

During the last conversation I had with Bill the dad on Friday, I instructed him to hold the pour until the following Monday morning to allow the trenches a little more time to dry. The next day, I went to the site to bail any remaining water out of the foundation trenches. Without notice to me, Bill the dad made the decision to place the concrete that day. I pulled up to the jobsite and saw a concrete pump at the curb with fresh concrete all over the public street.

The hill at the front of the property was all gumbo clay and as slick as a ski slope. The concrete truck could not get far enough up the slippery hill for its chutes to reach the trenches, so Bill the dad called for a high-pressure concrete pump. After pumping approximately half of a truckload, the pump malfunctioned and blew a large amount of concrete into the street. Bill and Bill had to finish placing the concrete by transferring it from the truck to the foundation with a front-end loader.

Situation—The concrete in the waiting truck had already been through more than 300 drum rotations. When that occurs, consolidation of the concrete starts to break down and the load has to be dumped. The driver wanted to dump the load (nine cubic yards) on my site or one of the vacant lots in the immediate area. The concrete company wanted full payment for the wasted concrete. The concrete pumping company wanted full payment for the aborted pumping service. If you were the general contractor, what would you have done?

My Decision—I informed the concrete truck driver that he could not dump the concrete on my site or on any vacant lot within the subdivision without the owner's permission. I refused to pay for the truckload of bad concrete, the concrete that was lost due to the pump malfunction and the cost of the concrete pump. Since Bill and Bill were responsible for placing and finishing the concrete, it was also their responsibility to pay for the materials and associated pumping service.

By noon, the concrete pump operator had his equipment cleaned and was ready to bid farewell.

Situation—I had a public street full of concrete and the person responsible for the mess preparing to leave the project. If you were the general contractor, what would you have done?

My Decision—I confronted the concrete pump operator and advised him that before he left my project, he had a street to clean. Rather reluctant and quite unhappy about his new assignment, the pump operator spent several hours hosing and scraping the semi-hardened concrete off the street surface.

If I did not go to the project that Saturday morning to bail water, I would have had a nine yard pile of dried concrete to move at my expense ($365), a bill for one additional truckload of concrete ($720), a bill for the concrete pumping service ($350) and several yards of semi-dried concrete, scattered around the public street that needed to be removed ($320). With everything taken into consideration, that trip to the site saved me approximately $1,755 in material and services that I would have been charged for, and several hours of time cleaning the street.

All of the concrete was placed and, thank goodness, got a good initial set before the clouds opened up again Saturday night.

FOUNDATION WALLS—Gumbo clay has the ability to hold water like a bathtub. With the tops of my footings four inches below grade, the trenches were full of water that resisted absorption into the ground. I spent most of Sunday digging relief trenches from areas around the foundation to the low side of the property to provide a positive drain for the water. That proved to be a smart decision, because it rained for most of the next five weeks.

On the days it didn't rain, I had Danny deliver and spread sand to stabilize the gumbo. The sand helped to absorb and drain the water that would otherwise be held by the clay. Danny also dumped several loads inside the perimeter of the footing to be used for fill material.

The subdivision was new at the time and had a dozen or more houses under construction. That was good for me, because I didn't have to travel out of my neighborhood to find subcontractors. In a short amount of time, I interviewed and hired "Efrain the Brick Man" as my masonry subcontractor, to install a concrete block foundation wall from the foundation to the first floor level.

I ordered a tandem (double) truckload of masonry sand and enough eight-inch concrete blocks and bags of mortar to build the foundation walls. I had the sand placed on the left side of the lot and the blocks and mortar placed to the right of the sand. My plan was to use as much sand as was necessary for mortar and, after Efrain was finished with the foundation wall, spread the remaining sand in the driveway to stabilize the gumbo.

Several days before Efrain was scheduled to start his work, we met on site to go over the plan. As soon as I pulled up to the site, I noticed half of the load of masonry sand was missing. It didn't take long to find it, however. My sand was lying in the driveway for the house to the left of my lot, which was also under construction. The workers, who were preparing the driveway for concrete, needed sand for the base and decided it was quicker and certainly cheaper to steal mine than to order a load.

Situation—Materials were stolen from my project and the culprits were obvious. If you were the general contractor, what would you have done?

My Decision—I confronted Gill, the owner of the construction company building the project next-door, in person, at his place of business. During our brief, but stimulating discussion, I voiced my displeasure with the actions of his workers and demanded restitution. Gill agreed to replace my sand.

The weather cleared and no rain was predicted for the next several days. Since mortar for masonry does not stick too well to gumbo residue and Efrain's crew was not interested in scraping and scouring the footings before they laid the block, it was my responsibility to clean all of the mud from the top surface of the concrete. It took one half of a day to clean the footings with a stiff brush and hose with a high-pressure nozzle.

Efrain started building the foundation walls on Monday, March 17 and completed the task the following day without any complications. After the foundation wall had several days to set-up, I had Danny load the interior with additional fill and compact the majority of the area with his front-end loader. To prevent the possibility of pushing the block out, the fill adjacent to the foundation wall had to be hand tamped. Two days later, and before the hand tamping was complete, however, it rained again. Following the route of least resistance, the water naturally flowed and collected in the areas containing the loose, uncompacted fill, causing soft mud. Although concrete blocks are porous and water will eventually seep through them, the foundation walls held the water and prevented the earth inside from drying.

Situation—The uncompacted fill adjacent to the foundation wall was saturated and the drying process threatened to substantially impede the progress of construction. If you were the general contractor, what would you have done?

My Decision—I drilled half-inch holes through the mortar joints for the first course of block, as close to the top of the foundation as possible. The holes, which were later plugged with mortar, acted as drains and allowed the water to escape. That action accelerated the drying process for the fill and allowed me to continue without any appreciable loss of time.

Approximately half of my sand remained after the masonry subcontractor completed the foundation wall. I decided to spread that sand in the driveway to mix with the gumbo and form a more stable base for better vehicle and walking traction. Danny was called in to perform that task. While Danny was busy spreading the sand, however, his front-end loader blew a hydraulic line, slid out of control and headed straight for my newly installed water spigot that I thought was well out of the way of the work area. There was nothing we could do but watch the tractor grind over the spigot and snap the water line in half. In addition, Danny knocked seven blocks loose while spreading the fill dirt. I closed the valve on the meter, repaired the water line, replaced the spigot then called Efrain. Late in the afternoon the following Wednesday, Efrain replaced the dislodged block.

Thursday, Brian, the owner of "Lights On Electric" installed a temporary service pole and called for its inspection. The electrical inspector approved the temporary service the following day. I established my electrical account, paid my meter deposit and on Tuesday, April 8, had electricity on the site.

Rain continued to hamper my efforts to get the fill area fully compacted and ready for the rough plumbing. Although water is used during the compacting process, fill material compacts best when it is damp. Excessive water converts the fill material to mud and does not compact well.

Late Saturday afternoon on April 12, I decided to check the site. When I arrived I noticed the cover for my temporary service panel hanging, the main breaker on the ground and a very long number ten wire hard wired to the supply side of my electrical service. Remember Gill? He was also building a second house directly behind the house next door to mine. Instead of installing a temporary service for each of the houses he was building, Gill installed one (for the first house) to serve both. Gill's flooring subcontractor was installing and sanding wood flooring in the second house and needed 220-volt current to operate the floor sander. Although my temporary service pole was approximately 250 feet from where they were working, the workers walked about 70 feet past Gill's temporary service and tapped into mine. I disconnected the wire, replaced the breaker and installed a rather large combination lock on the panel door.

Sunday, I went to the site early in the morning to do some hand tamping, and discovered that the trunk sections of three very large trees appeared on my site since the evening before. Several lots had been recently cleared in the immediate area, therefore, I could not determine the origin of the trees. The trees were blocking the limited access I had for material delivery, so two days after they were left on my site, I prepared to cut them into smaller sections and have them hauled off. When I arrived on the jobsite with my chain saw in hand, however, the trees were gone. I do not know how they got there and I do not know how they were removed.

PLUMBING STARTED—I hired "Peter Plumber" as my plumbing subcontractor. Peter and I made an appointment to meet on site to go over the plumbing plan. The very first thing Peter stated was that he could not work from a conventionally drawn plumbing plan (which provides rough-in dimensions to the center of the pipe inlets and outlets). He said the plan needed to provide the dimensions to the edges of the inlets and outlets instead. All of the dimensions had to be changed to suit Peter's rough-in style. I made the changes and we agreed that Peter would start on Monday, April 21.

When Peter did not show up Monday morning, I called him on his cell phone. Peter informed me that he was on a weekend fishing trip and was sick. He said he needed one day to recover and would be on the job bright and early Tuesday morning. Tuesday came and went and Peter was neither bright nor early. In fact Peter did not show up at all. I finally got in touch with him Tuesday evening and learned that he was selected for one day of jury duty for magistrate's court. Then it started to rain.

Peter started his rough-in that Friday, completed his work the following Tuesday and called for its inspection. The rough-in passed inspection and we were ready to backfill the pipe trenches. The house had an open floor plan with the kitchen sink in the middle of a counter. Behind the sink was an open wall and a raised dining bar. The revised plumbing plan was very easy to read. Yet, when I checked the actual layout after the rough-in, I noticed that the drain and water supply piping for the kitchen sink was stubbed nine feet short of where it was supposed to be. If not corrected before the concrete for the floor was placed the water and drain piping for the kitchen sink would be in the guest bedroom. Naturally Peter was surprised to learn of the mistake, but corrected it the next day. Nonetheless, I still lost another precious day in time.

Peter used a small tractor to backfill the trenches. During the backfill process, he knocked two blocks out of the garage foundation wall. In addition, he used wet gumbo to backfill his trenches instead of the sand that was provided. The blocks had to be reset and the gumbo had to be removed, the trenches refilled with sand and tamped by hand. "Bug Off Exterminators" was scheduled to perform the termite treatment late in the afternoon on Tuesday, May 5, and the concrete due to be placed the next morning by "Almost Level Concrete Company." I had to call both subcontractors and delay their tasks for the two days it took to perform the repair work.

THE FLOOR SLAB—Thursday, May 8, I took a day of annual leave to clean the fill that the rainwater washed over the interior footings and prepare the slab area for the termite treatment. I scheduled Bug Off two days in advance. John showed up on time and as soon as he saturated a section of the filled area with chemical treatment, I installed a polyethylene vapor barrier over its surface. Quick installation of the vapor barrier over the treated fill prevents evaporation of the chemical treatment and assures maximum protection against termites.

Earlier in the day I called the building inspection department and requested my slab inspection be performed late in the afternoon. As I just completed the installation of the vapor barrier, the building inspector arrived. As soon as the building inspector passed the inspection, I called Mike, the owner of Almost Level, and verified that he would have his finishing crew on site Friday morning. Mike said that his crew would be there, but he was preparing to leave for Myrtle Beach for a long weekend. Since Mike was the crew chief, I voiced my concern about him being on vacation while his crew was working on my project. He informed me that his nephew worked with him for several years and was very capable of running the crew. To raise my level of comfort, I got Uncle Mike's cell phone number.

On Friday, nephew and his finishing crew showed up on time and started their prep work. When I saw the first concrete truck roll up to the site on time, I knew that I would have a floor by Friday afternoon.

Although my house was being constructed 180 miles from the South Carolina coast, I still designed it to withstand coastal, hurricane force winds. In an effort to get a framing subcontractor with the knowledge and qualifications to build to the more restrictive hurricane standards, I hired "Good Brother's House Framing Company" from

Myrtle Beach. When Almost Level started placing concrete, I called Kenny, the crew chief for Good Brother's, to confirm that his crew would start the wood framing Monday morning. Kenny said the schedule was set and he would see me at the crack of dawn.

With the framing subcontractor confirmed to start, I had to leave the project to order and arrange delivery of enough framing lumber to last a minimum of five working days. I informed nephew that I had to leave the site for several hours. Before I left, however, I gave nephew 95 foundation bolts and informed him twice, that the bolts needed to be set for standard two inch by four inch exterior stud walls.

I left the project and headed to the lumber supplier, placed my order for my framing materials and requested the last delivery of the day to keep the material out of the way of Almost Level's crew. Since my plan was designed with 10-foot ceiling heights for all rooms on the first story, I called the lumber supplier several days in advance to confirm that 10-foot studs were available. At the time I placed the order and before I left the lumber supplier, I reconfirmed availability of the studs.

I was away from the project for almost four hours. By the time I got back to the site, nephew was gone and several members of his crew were preparing to leave, but three workers remained an additional half hour to complete the surface finishing. The workers looked like they were doing a good job, so I decided to clean the dirt and trash that found its way from my jobsite to the public street.

Concrete increases in strength, hardness and has a much higher resistance to developing (shrinkage) cracks if it is kept moist after its initial set. About one hour after the last three members of nephew's crew left, I connected my hose and prepared to dampen the new concrete. As I got closer to the floor, I noticed that several anchor bolts were placed too far inward from the exterior of the block. A quick examination of the perimeter of the slab revealed 41 anchor bolts that were improperly installed. I placed a call to Uncle Mike's cell phone and interrupted his vacation in Myrtle Beach. I informed uncle that he needed to contact his nephew and get him back to the job to remove and reset the bolts while the concrete was still "green." Nephew did not return to the site.

At approximately 5:30 that afternoon, my framing material was delivered. The load looked a little skimpy, so I performed an inventory and discovered the wall studs were missing. I asked the delivery truck driver why the studs were not loaded with the other materials. The driver informed me that the material lumberyard was out of 10-foot studs. Although I knew the lumberyard closed at 4:00 PM, I called anyway. The person that answered the telephone informed me that everyone was gone for the day and would not return until the following Monday. I asked for my salesman's home telephone number and was informed that it was unlisted.

Situation—I had a crew traveling 180 miles to start the framing for my house the next working day. The first components they would need were paid for, but not on the job and the material supplier had none in stock. If you were the general contractor, what would you have done?

My Decision—I called the store manager at home, explained my dilemma in a calm manner, but in painful detail and reminded him that he took my money but did not produce the product. I suggested that he contact another store, locate the studs and deliver them to my jobsite ASAP. He called me back in 15 minutes and informed me that the studs would be on my project Monday morning, by the time I got there. They were.

Situation—I had 41 out of 95 anchor bolts that were completely nonfunctional and would actually be on the room side of the finished, exterior walls. The concrete was set and there was no hope of removing and resetting the existing bolts. All 41 anchor bolts had to be cut out and new bolts installed. If you were the general contractor, what would you have done?

My Decision—Saturday, May 10, I rented a hammer drill with a one inch masonry bit, purchased 41 anchor bolts, a bag of high strength, quick setting cement and headed to the site to drill holes and set bolts. All related costs to replace the bolts, including my time, were deducted from my payment to Almost Level Concrete Company.

On Sunday, (Mothers Day) I went to the site to check the set on the new bolts and cut the useless ones out. Before I actually started work, and very much to my surprise, nephew showed up and asked what he could do to help. I told him he could cut the 41 useless bolts out and grind several rough areas of the concrete. We both spent the majority of the day trying to clean-up sloppy sections of the hardened concrete before the framing crew arrived. When I got home later that afternoon, I had a telephone message from Kenny. He had to be in traffic court, in Myrtle Beach on Monday morning. I lost another day. Tuesday, May 14, Kenny and his framing crew showed up and started working.

Midmorning on Tuesday, I took two hours of annual leave to drive to ReddiTruss, a truss manufacturing company, to order my roof trusses and two very long beams that had to be custom made to span the openings at the front and rear porches. Beams are usually attached to wall systems with components known s "beam hangers." When I ordered the beams, I also told my sales person that I needed four beam hangers as well. The beam and hangers for the front porch did not present any problems. The beam for the rear porch, however, was very long and had to be connected to an exterior wall that had a 90-degree angle on one end and a 45-degree angle at the other. Those beam hangers did not exist.

Situation—I had beams that were special ordered that needed to be connected to the wall system with beam hangers that did not exist. If you were the general contractor, what would you have done?

My Decision—I called the engineering department for the largest wood frame connection manufacturer in the country, explained my circumstance to one of its design engineers and asked if the company could manufacturer two special beam hangers. The answer was yes, with an approximate delivery time of six days.

THE CRANE INCIDENT—While I was waiting for the trusses to be manufactured, I called several crane companies to get prices and to check on availability of equipment. The limited working space on my site would require the crane to be setup and function from the street. In addition, due to the limited area I had on my site to store materials, I also had to have the crane to off load the trusses from the truck and place them close to the house. My roof trusses were 56 feet in length. Trusses that large, coupled with the distance they needed to be moved required a crane with a 50-ton load capacity and a very long reach. Only one crane company within a reasonable distance had a crane with those specifications. That narrowed my search (and the competitive pricing) dramatically.

A roof truss is a component that combines a roof rafter and ceiling joist into a single element. The rafters and joists are braced with interior supports referred to as webs. The rafters, joists and webs are connected with gussets, which are metal plates with sharp prongs that are pressed into the wood members under very high pressure, at each joint. Roof trusses are very strong components once they are set on the building walls and properly braced. While they are being moved into place, however, roof trusses tend to be flimsy and vulnerable to damage. When setting with a crane, a "spreader" (steel brace) is usually necessary to keep the trusses from bending at the center gussets after they are lifted off the ground.

I hired "Green's Crane Company" and met with its representative, (another Mike) on site. During our meeting, we agreed that a spreader would be necessary. We scheduled the crane with a spreader to be ready to off load the trusses when the delivery truck arrived at 8:00 AM, Monday, June 16.

On the Sunday before, however, the rain started again and continued into Monday. The trusses could not be delivered due to the weather, so ReddiTruss rescheduled delivery for 8:00 Tuesday morning. I called Green's and advised Mike of the change of schedule. Mike told me that the rain affected all of Green's schedules as well, but quickly assured me that the crane would be there when I needed it.

The trusses were delivered to the site on Tuesday at 9:30 AM, but the crane was not there to off-load them from the truck. I called Green's and was told that Mike had to go out of town and no one else knew about the crane for my project. Meanwhile, Eddie, the dispatcher with ReddiTruss was holding for me. Eddie said, with semi panic in has voice, the truck that was waiting to be off-loaded at my site was needed for other deliveries. After I calmed Eddie down, I called Green's again. This time I spoke with Earl in dispatch and stressed the urgency of getting the crane to my site. Earl said Mike did not pass the information about my order to him. I quickly went through my discussions with Mike and gave Earl the specifications for the crane I needed. Earl informed me that all of Green's 50-ton cranes were out on other jobs. He said that the earliest he could have a crane on my site was later that day.

Eddie called me back and was frantic. In a loud voice with elevated tones, he advised me that I was responsible for holding up the framing crews for several other jobs. He said he was going to tell the truck driver to dump the trusses in the street. I asked him to give me five more minutes.

Situation—A flat bed truck with an order of roof trusses for my project was waiting at my construction site to be off-loaded. To make matters worse the dispatcher for the truss company who was in full panic mode was threatening to dump my trusses in the public street. If you were the general contractor, what would you have done?

My Decision—I called Kenny and told him to have the driver drop the trusses on a vacant lot that was diagonally in front of my project.

My construction site was located 12 miles from my office. From a time prospective and under normal conditions, I could go from my desk to my project in 20 minutes. Fifteen minutes after I told Kenny to drop the trusses on the vacant lot, I was there, surveying the pile. The trusses slid off the truck in a single pile and were not damaged. That was good.

In the process of maneuvering the truck to drop the trusses, however, the driver pushed my portable toilet in front of my neighbor's driveway. The truck also rolled over the curb and created large tire ruts in my neighbor's front yard and broke the four-inch sewer inlet that was serving my lot. Kenny and I pushed the portable toilet back to its original position in front of my site.

Situation—A truck that delivered material to my project damaged my neighbor's front yard. The ruts needed to be filled and tamped, the damaged area regraded and new sod installed to restore the yard to its original condition. If you were the general contractor, what would you have done?

My Decision—I charged off the balance of the day to annual leave and personally repaired my neighbors yard as well as my sewer pipe.

It was already Tuesday afternoon and the crane was nowhere in sight. I called Green's again and spoke with Ryan. He informed me that he was not familiar with my order, but would check with dispatch and call me back in five minutes or less. I went back to work on my neighbors yard and lost track of the time. Two hours later, I realized that Ryan had not called me back, so I called him. Ryan informed me that the crane scheduled for my site had to have about one hour of work performed on it before dispatch would send it out. One hour later, I called Ryan again. Ryan said that the crane was not ready and suggested we reschedule the delivery for 8:00 Wednesday morning. The day was already shot, so I agreed. I also reminded Ryan that I needed a spreader. He said "no problem" as he hung up the telephone.

Before I did anything else Wednesday morning, I spoke with Kenny by telephone. He informed me that one member of his framing crew did not show up for work.

Situation—Setting trusses requires a minimum crew of four workers. If you were the general contractor, what would you have done?

My Decision—I called my office and took another full day of annual leave to help the crew set trusses.

I got to the site a few minutes after 8:00 AM. No crane. Before my truck came to a full stop, I was already dialing Green's telephone number. Ryan informed me that the crane scheduled for my site was sent to handle an emergency that developed on another project. He said that dispatch would send a crane to my site ASAP. Since I had a crew standing around, doing nothing while they were waiting for the crane, I pressed Ryan for a specific time. He said he would have to speak with the crane operator before he could give me a specific time and asked me to call him in 15 minutes.

Exactly 15 minutes later, I had Ryan on the telephone. He informed me that he did not have the opportunity to speak to the crane operator, but he would definitely do it within the next 30 minutes. It was obvious that Green's had more work than it could effectively handle at that time. Under normal circumstances, I would have dropped Green's after the third telephone call, but being the only supplier in the area that had the equipment I needed, I was stick with it.

Over the next several hours, I called Green's eight more times. During the eighth call, Ryan finally admitted that Green's could not supply a crane on Wednesday. He guaranteed the crane would be on the site 8:00 Thursday morning. Shortly after noon, I dismissed the framing crew for the day. Later in the day, a representative of Green's showed up on site to see how the crane had to be setup. We both agreed that the crane had to be set in the middle of the street. Since the trusses were long and had to be moved over the street, we also agreed that a spreader was necessary.

The full Good Brothers crew and the crane were on site by 8:00 Thursday morning. I stayed on the job long enough to be assured that the crane setup and crew preparations were progressing in a normal manner, and then went on to my office. After work, I headed to the site to check the progress. As I pulled up to the project, I was surprised to see the crew still there. I quickly learned that the last three trusses to be set, cracked at their centers while the crane was moving them over the street. When I inquired why the trusses cracked, I was informed that the trusses were being moved without the spreader that Green's promised, but failed to provide.

Situation—Three damaged trusses brought all construction progress to a standstill. The framing crew and crane was sitting idle. If you were the general contractor, what would you have done?

My Decision—I told Kenny to make temporary repairs to keep the damaged trusses stable and allow the completion of the set.

The time required to make a temporary repairs further delayed the truss setting process. I was forced to keep the crane in place and finish the set on Friday. With all of the trusses set and braced, I concentrated on a permanent repair for the three trusses.

Situation—Trusses are engineered components. Any permanent alterations or repairs made to a truss (or any engineered component) have to be designed or approved by a structural engineer. If you were the general contractor, what would you have done?

My Decision—I contacted the design engineer for the truss manufacturer, and also Greg, an independent structural engineer familiar with correction of truss damage, and requested a repair that would equal or exceed the strength of the original design.

Both engineers agreed on a method of repair, which I followed precisely.

The time I spent working with the Good Brothers framing crew allowed me to evaluate its workers. It was my conclusion that one of the workers was not experienced enough to perform carpentry tasks. I advised Kenny that I did not want the person back on the job. Monday, June 23, Good Brothers arrived with three workers, including Kenny.

Situation—To keep the project moving, it was necessary to get the roof sheathing and felt installed before the next rain. With a roof as large and steep as mine, it was necessary to have a minimum of four experienced carpenters to safely install the sheathing. If you were the general contractor, what would you have done?

My Decision—Hire an additional framing crew to install the sheathing and felt while Good Brothers continued to work on bracing and additional framing in other areas of the house.

THE SEARCH WAS ON—I wanted four experienced workers. That meant I had to find a whole framing crew that could work me into its schedule. My adventure started Monday morning, June 23, with phone calls to several of the larger framing subcontractors in the area. All had as much work as they could handle. By chance while I was at a lumber supply store later that day, I ran into Rick, an acquaintance who happened to be in charge of several framing crews. I explained my dilemma and asked if he could spare four carpenters. He said he would check and call me the next day. Rick called me early Tuesday morning and informed me that Jerry, the lead carpenter for his best crew, would be on my jobsite with three additional carpenters Wednesday morning. Rick gave me Jerry's cell phone number and said I could deal with him directly, from that point forward.

Wednesday morning Jerry and his crew did not show up on the job. When I called him on his cell phone, he said he had to go to hospital Tuesday night and was up quite late. He said he would have his crew on the job Thursday morning. Thursday, neither Jerry nor his crew showed up. I called Jerry and informed him that I would not call him again. Friday Jerry showed up late in the day with one helper. Neither was prepared to work. He said he would be on the job Monday for sure. At that point, I informed Jerry that if he and his crew were not there Monday, I would get another crew to perform his work. To make up for lost time, Kenny told me the Good Brothers crew would work Saturday and Sunday. One member of his crew, however, refused to work weekends and I would have to work in his place. Being somewhat behind schedule, I took him up on his offer.

Saturday was a short day due to rain in afternoon. As we worked, one of the Good Brothers told me he had a friend in Columbia who was a good carpenter, and that he would be willing to work on Sunday. He said if I were interested in hiring him, he would make the contact. I was interested. Later Saturday evening I spoke with my prospective carpenter by telephone. We agreed on terms and I hired him for Sunday. Saturday night the Good Brothers crewmember that gave me the tip and his buddy the carpenter, went out on the town. They had either a good night or a bad night, depending how you look at it, and neither showed up for work the next day. Sunday Kenny and I worked 11 hours. Late in the day the trigger on the pneumatic nail gun malfunctioned.

Monday, June 30, I dropped the nail gun off at the tool repair shop then stopped by the project. All of the Good Brothers as well as Jerry and his crew showed up ready to work. Jerry and his crew left early due to an afternoon shower. The Good Brothers crew stayed, but the crewmember that partied so hard on Saturday night must have been having flashbacks. He could not concentrate and kept making mistakes. When Kenny questioned him about his situation, he got mad and quit. Tuesday, the rain returned and lasted three more days.

A CONFRONTATION—The crews returned to work on Monday, July 7. Jerry and his crew, however, worked sporadically and did not keep consistent hours. Friday was payday. When I got the hours that each person worked from Kenny, I noticed that the invoice submitted to me by Jerry was substantially out of proportion. Although in our contract we agreed that the crew would be paid for each hour worked, Jerry charged me a daily rate based on eight hours per day per worker. His invoice, therefore, reflected a full eight-hour day for each worker just for showing up, regardless of the hours worked.

Situation—I was charged a daily rate of eight-hours per day, for three people, regardless of the number of actual hours worked, instead of by the hour as agreed. If you were the general contractor, what would you have done?

My Decision—I recalculated the amount of the invoice based on actual hours worked and prepared the checks accordingly.

In addition, Kenny informed me that he suspected Jerry and his crew of taking marijuana breaks. Kenny informed Jerry that I would not tolerate drug use on my project. Although he did not admit to smoking marijuana on the job, Jerry quickly snapped back by declaring (in his words) "what I do in my truck is my business."

Situation—I had three workers smoking dope on my construction project. If you were the general contractor, what would you have done?

My Decision—I confronted Jerry and his crew and informed them that what they did on my project, whether inside a truck or not, was my business as well as theirs. I gave them a brief but stern lecture and then fired the entire crew.

One of the fired workers had an issue with the fact that I cut his time to hours actually worked and approached me. After he blew off a little steam he said that "the Lord" was going to punish me! I asked him if he was making a threat against my job or me. He said no but repeated the comment. I asked him a second time if he was making a threat. Again said no and went on to recite what he thought were quotes from the bible. Listening to all the gospel according to pothead that I was in the mood to hear, I told Jerry and his crew to leave and not return.

Note! I do not care what the people I hire do after work hours and away from my project. That is truly their business. I have a strict no dope policy on my projects, however, and violation warrants swift action on my part. In addition to being illegal, dope on your project is a major danger to all workers, which could result in injury or death. Insurance and workman's compensation will not cover dope related injuries. As the owner/contractor, you have no way of knowing individual or collective work habits or ethics of members of the various sub crews. The experienced contractors learn who the undesirable workers are through trial and error and word of mouth. Checking references is critical, but as in my situation, certainly not foolproof.

THE SEARCH WAS ON AGAIN—Saturday, I looked for a framing crew to supplement the two remaining Good Brothers. No luck. Monday, July 14, I took two hours of annual leave to scout jobsites. I spoke with several crew chiefs, but still no luck. Late in the day on Wednesday, I got a call from one of the framing subcontractors I had contacted. He informed me that he would have a crew of six workers on my job Thursday morning. Wednesday evening, Kenny's remaining worker called me at home and informed me that he would not be at work Thursday morning. That reduced my original framing crew of four, to one. Since I had a crew of six due on the job, I would let Kenny work with them.

To finish the roof sheathing, I needed one more load of plywood on the site when the crew arrived. Thursday I rolled out of bed at 5:45 AM and went to buy the plywood. When I placed the order, I was informed that the price increased by two dollars a sheet over my initial quote for plywood and oriented strand board (OSB). Vigorous arguing with my sales representative got me nothing more than frustrated. If

I wanted the material, I had to pay the additional cost. In an effort to get the plywood to the job as quickly as possible, I had it loaded in my truck and I delivered it personally. Kenny helped me unload the wood and I went on to work.

By the time I got to my office, I had several job related calls on my voice mail: the nail gun was repaired and I could pick it up at any time; the millwork shop needed payment for the interior doors; dry wall and painting subcontractors wanted to meet with me and offer quotes. Each call required my physical presence. At that point, I decided to take the rest of day off to take care of those chores and do miscellaneous work on site. When I arrived on site, Kenny was there, but my crew was not. Kenny and I worked 10 hours. Friday still no crew. I decided to take another day of annual leave. Kenny and I worked nine hours.

Friday afternoon I called another framing subcontractor, but to no avail. When I got home, later that evening, the Good Brothers crewmember that quit was on my answering machine. He informed me that he was working with another crew, but was willing to work weekends. I told him to be on job 6:30 the next morning. Saturday the crewmember showed and he, Kenny and I worked all day. Sunday I worked all day by myself, installing bracing.

Monday, July 21, Kenny returned with an additional crewmember, bringing the Good Brothers crew up to three. Intermittent rain Monday and Tuesday slowed progress. Wednesday evening I met with another framing crew chief on the site. He promised to be on the job the next morning with one additional carpenter.

Thursday, the Good Brothers crew, including the member who quit, as well as the other crew chief with his one carpenter showed up. The six workers made so much progress that they ran out of nails, framing lumber and house wrap. I took two hours of annual leave to get and deliver the materials to job. The only components necessary too completely dry-in the house were the windows and exterior doors. When I called my supplier to schedule delivery of the windows and doors for the next morning, I was informed that only one truck was in service and it was already scheduled for entire day. My windows and doors could not be delivered before Monday.

Situation—We were on a roll and I wanted to have the house dried-in by Sunday. If you were the general contractor, what would you have done?

My Decision—I rented a truck, took a half-day of annual leave, picked up the windows and doors and delivered them to the site.

DRYING-IN—Saturday I worked with the Good Brothers to get the house dried-in. Monday, July 28, I took two hours of annual leave to return 16 sheets of plywood and five rolls of roofing felt for credit. After work on Tuesday I met on site with Gene, the owner of "Lotta Air" Heating and Air Conditioning.

Wednesday I took one and a half hours of annual leave to select and purchase my bathtubs, shower units and water closets. I ordered a solid acrylic, single-piece tub/shower combination unit for one bathroom and a single piece shower unit for the other and a whirlpool for the master bath. The bathroom and shower units were ordered with dome lights. After work on Wednesday, Joan and I met Brian from Lights On Electric to do a walk-through to mark the locations of our switches, receptacles and lights. Joan and I spent two hours with Brian then we were finished for the night.

It rained on and off (mostly on) for next week. Except for the electrical work, very little progress was made. After work on Wednesday, August 6, I met with Efrain the Brick Man. Efrain committed to continue the brickwork the following Monday. Based on that commitment, I ordered my brick, mortar and sand and scheduled delivery for Friday. Saturday I got out early to shop for a gas fireplace. By noon I was on the job and I worked until sundown.

On Sunday I worked all day installing additional roof bracing and "cat walks" in the attic.

Monday, August 11, Efrain's crew did not show up. I called Efrain and he assured me that the crew would be on the job Tuesday morning. I also called the plumbing supply store to check on my bathtubs and shower unit. They had arrived (the dispatcher failed to call me to schedule delivery). We scheduled delivery for Tuesday. As soon as I confirmed the delivery, I called Peter Plumber and Lights On Electric to schedule their crews for Tuesday morning.

PLUMBING PROBLEMS—On Tuesday Efrain's crew, Brian's crew and Peter's crew showed up on time. Later Tuesday morning, the tub/shower and shower units arrived on site and Peter's crew installed and connected them. When I arrived on site after work, Peter was still there. As soon as I saw the tub/shower unit, I realized it was not what I ordered. In addition, neither the tub/shower unit nor shower unit had the dome lights installed.

Situation—The plumbing subcontractor unknowingly installed and connected the wrong tub/shower unit. If you were the general contractor, what would you have done?

My Decision—I called the plumbing supplier and had the unit model and my order verified. I then instructed Peter to remove the tub/shower and I placed a new order for the correct unit. Since the supplier made the mistake, I insisted that Peter bill the supplier for his crew's time to install and remove the tub/shower.

Situation—The shower unit was installed and all piping connected without the specified dome light. If you were the general contractor, what would you have done?

My Decision—Rather than disconnect and remove the shower unit, I decided to have the manufacturer supply the light kit and I would ask Brian to install it.

Wednesday, August 20, I hired "House Topper Roofing" to install the shingles for my roof. Ron, the owner of House Topper said he could start the following Friday. Friday arrived and House Topper was not topping my house. When I called Ron to inquire why he did not start, he said he was finishing another house and would have his crew on my roof Monday for sure.

ELECTRICAL PROBLEMS—Saturday I went to the house to work. During the course of the day, I inspected the wiring that was installed to that point. I noticed 14-gage wire was installed for many of the light circuits. Although 14-gage wire is acceptable for lighting circuits, I prefer and always specify 12-gage (thicker) wire.

Situation—My electrical subcontractor installed thinner, but still acceptable gage wire, contrary to my electrical specifications. If you were the general contractor, what would you have done?

My Decision—I called Brian and informed him that the 14-gage wire must be removed and replaced with 12-gage wire.

The lender that I used for my construction loan required a plot plan, showing and verifying the location of my foundation, before dispersing the first draw. Monday, August 25, I ordered a plot plan from the original land surveyor. The office administrator explained that the field crews were quite busy, but anticipated the plot plan could be produced in "several" days.

MORE PLUMBING PROBLEMS—On Wednesday the plumbing supplier received the reordered tub/shower. We scheduled delivery for the next day, and I immediately called Peter. Thursday Peter set and connected the new tub/shower. Later that day, I inspected the installation and noticed a large flaw on the front surface and the unit itself was not square in the framed opening. I verified that the opening was square and the floor was level.

Situation—After it was installed and all piping connected, the second tub/shower unit was out of square with the framed opening and contained a major flaw in a very obvious location. If you were the general contractor, what would you have done?

My Decision—I stopped the plumbing subcontractor from going any further, called the plumbing supplier, advised my salesperson of the problems and requested a factory representative visit my site and schedule an onsite repair by a factory technician.

As I continued to inspect the area around the tub/shower, I noticed that Peter accidentally cut a high voltage wire while in the process of installing the vent piping through a top plate in the bathroom. I marked the location of the damaged wire and called Brian to replace it.

Tuesday, September 2, I took two hours of annual leave to meet with Gene to go over the HVAC equipment locations and duct layout. I also went to the land surveyor's office to inquire about my plot plan. The office administrator advised me that all crews were very busy, but my project would be "worked in" within the next day or so. Later Tuesday, I called the manufacturer of the tub/shower, to get a work order number for its repair. The helpful representative informed me that I could not get a work order for the repair without the serial number for the unit. She advised me to get the serial number and call her back the next day.

Wednesday Lotta Air started work. About midmorning I called the tub/shower manufacturer with the unit serial number and got a work order number for its repair. Thursday Lotta Air completed its rough-in. I inspected the work and discovered several duct joints were not sealed with mastic. When I informed Gene about the discrepancies, he said he would correct them the next day.

Monday, September 8, the tub/shower factory representative called me and said he would be at my house later that day to look at the unit. I took two hours annual leave and met with the representative. He informed me that the tub/shower could not be repaired in any manner that would square it in the opening or completely remove the flaw. He then said that he was authorized to offer me $175.00 if I would accept the unit and "live with" the flaws. The only other option was to replace the unit a second time.

Situation—My tub/shower could not be repaired to look like it was a first quality run, but I could accept a payoff to live with it. If you were the general contractor, what would you have done?

My Decision—Accepting cash to compensate for a flaw in a component is always an attractive option. In this instance, however, the flaw was on the front of the left side directly at eye level and the degree that the unit was out of square was very noticeable. To live with it meant that my guests and I would see the obvious flaws every time we occupied the bathroom. I told the factory representative to replace the unit.

Tuesday the Lights On crew was back on the job. I took two and a half hours of annual leave to meet at the house with prospective insulation subcontractors. Later Tuesday I called Marc the sales representative for "Hot Stuff" Fireplaces and made an appointment for the following Thursday. Tuesday evening Joan and I met with one of several cabinet subcontractors.

ROUGH-IN INSPECTIONS—Monday, September 15, I went to the land surveyor's office and got my plot plan. My goal for the week was to get my framing and rough-in inspections by Friday. I had my alarm subcontractor "Burglar Be Gone" and Peter Plumber scheduled to finish by Thursday, and Lights On and "Long John's" Insulation Company by Friday.

Thursday Brian's crew worked half of the day. The crews for Burglar Be Gone and Peter Plumber did not show up at all. Friday Brian's crew was back was on the job. When I called Jamie, the owner of Burglar Be Gone, I was advised that the crew was on its way to my project as we were speaking. Shortly after noon I had an appointment on site with Shawn, the co-owner of "Top Dollar" Cabinets. When I got there, however, the plumbing and alarm subcontractors were nowhere to be found. I called Peter and Jamie and was assured by both that their rough-ins would be complete by Monday. Since the insulation has to go in after the wiring and piping, I had to schedule Long John's for following Tuesday. The appointment and phone calls cost me another three hours of annual leave.

Monday, September 22, my electrical system was ready for inspection, the alarm system was started and Peter's crew was still lost. Peter apologized profusely when I called him and again assured me that the plumbing rough-in would be completed Tuesday. I called Craig, the crew chief for Long John's Insulation Company and rescheduled for Wednesday.

The job always ran smother when I was physically present, so I took a full day of annual leave on Tuesday to get the house ready for the required inspections. Peter finished the plumbing system and performed the pressure tests. Mid morning I took the excess materials back for credit. Later in the morning I met Larry, the owner of "Lumpy Larry's" Paint and Drywall on site. After lunch I performed a thorough inspection of the electrical and plumbing systems. During the inspection, I noticed the box for one ceiling light was not installed according to the plan, the valve box for the washing machine was set in the wrong location and a test plug for the drain, waste and vent system had blown out. I called Brian and Peter and asked them to correct the discrepancies by Wednesday afternoon. Late in the day I called the Richland County Building Inspection Department and scheduled my framing and rough-in inspections for Thursday. Wednesday Jamie's crew finished the alarm system, Long John's finally started work, Peter reset the washing machine valve box and test plug, and refilled the system with water and Brian installed the forgotten light box.

Thursday Long John's had about two more hours worth of work to perform to be finished when the building inspector, the only person to show up early during the entire length of the project, rolled to a stop in front of my house. The inspector saw that the insulation was not ready and would not perform any inspections. Later in the day I spoke with Marc by telephone. He said he would meet me on site 8:30 Friday morning. I also started searching for a gas subcontractor to install my gas piping.

Friday Marc did not show up. To give him the benefit of the doubt, I waited until 8:50 AM then gave up and went on to work. About 9:30 AM, Marc called my office to inform me he was at the house. We established the location for the fireplace by telephone. He said he would calculate the total installed cost and e-mail a firm price by the end of the day. He didn't. I called the building inspection department and requested a reinspection for the following Monday.

The fireplace quote was received on Monday, September 29. I accepted the price and was informed that the fireplace would be delivered the following Thursday and installed Friday. Later in the day, the building inspector passed all systems. That evening, I called Lumpy Larry and scheduled the drywall installation to start Wednesday morning.

MORE ADVENTURES—Wednesday Larry's crew failed to show up. Marc called and informed me that the fireplace would be delivered October 8 and installed the following day. Gene from Lotta Air told me that his company was licensed to install gas pipe. He also said if I gave him the job, the piping would be installed "in the next day or so."

My plan called for black pipe to be used as the piping material for the gas system. When working with black pipe, the lengths are cut to size and threaded on the job. That Friday, both Joan and I took a day of annual leave to shop for flooring and other finish materials. Saturday the drywall hanging was complete. During my inspection of the work, I noticed the boards did not contain enough screws in the centers and several electrical wires and boxes were covered.

Monday morning, October 6, I called Gene to schedule the installation of the gas piping. Gene then advised me that he did not have a pipe-threading machine and preferred not to do the job. I also met with Larry on site to discuss the drywall discrepancies. He agreed to make the corrections. Later in the day, I spoke with Frank, the owner of "Life's a Gas" and made an appointment to meet him on site the following Friday to review my gas piping needs. Tuesday the drywall finishing started.

Wednesday I took two hours of annual leave to go to the brick plant to order and pay for more brick and mortar. Thursday Hot Stuff did not show. I called Marc and left word on his voicemail. Marc did not return my call. Later in the morning, I took two hours of annual leave to meet Danny, my grading subcontractor on site. Thursday afternoon I called Life's a Gas to confirm my Friday morning meeting with its estimator. Friday I arrived on site, but did not see the estimator. When I called to inquire about the appointment, the dispatcher informed me that she attempted to contact me by telephone earlier that morning to reconfirm the time. She said since she could not reach me, the estimator assumed that I was not going to be available the rest of the day and cancelled the appointment. She also informed me that the estimator would not be able to meet me until the following Tuesday. The estimator's decision to cancel our meeting cost me one hour of annual leave. Later Friday morning I sent Marc an e-mail requesting that he call me as soon as possible.

Monday, October 13, Marc finally called and told me that the fireplace was in, but he would have to call me back to schedule installation. Later that morning I called Efrain inquiring about his schedule. He informed me that his crews were backed up due to four days of rain the prior week. He said he would try to continue the brickwork on Wednesday or Thursday. Tuesday the estimator for Life's a Gas showed up for our appointment. We agreed on a price and scheduled installation for Friday. As I was going back to my truck, I noticed that my portable toilet was missing. While I was in the process of calling the rental company, I saw one of its service trucks in the area and informed the service technician of the obvious prank. We found my toilet unit one block away lying on its side. I helped the technician get the unit back to my jobsite then headed back to my office. The total amount of spent annual leave amounted to two and a half hours.

Thursday I noticed a wet area on the floor in my rear bathroom. After I investigated the situation, I concluded that a drywall screw was driven into a water line and promptly called Peter to replace the damaged section of pipe. On Friday Efrain's crew showed up, Life's a Gas did not. I called Frank and was informed that his crew may start the following Monday. Tuesday, October 21, the gas piping was installed and inspected the next day.

Thursday I took two hours of annual leave to get brick lintels for the masons. Wednesday Efrain informed me that he needed 10 more cubes of brick (5,300 brick) to finish. After I questioned Efrain's calculation, he quickly reduced the number to eight cubes. Still believing the number was high, I ran the calculation myself and determined that 3,000 brick (six cubes) would complete the job. When I called the brick plant to place the order, my sales representative informed me that the brick had to be shipped from another plant (96 miles away) and the delivery date would be November 3. As much as I hated to do it, I had to send the masons home.

The brick was delivered that Friday. Tuesday, November 4, I stopped by the job before going to my office. Efrain informed me that he needed 10 more bags of mortar. When I got to my office, I called my sales representative and placed the order. After the order was placed, I requested an approximate time for delivery. I was told that deliveries were behind schedule and the mortar could not be delivered for several days. I took two hours of annual leave to pick up the mortar and deliver it to the job. Later that day the drywall finishing was completed.

Saturday the Good Brothers started the interior trim work. Sunday, while painting in the kitchen, I noticed that an electrical receptacle scheduled to be placed on the breakfast nook side of the wall was installed in the space reserved for the dishwasher. I also noticed the wire for the "glass break" detector for the alarm system in the master bathroom was installed where a cabinet was to be placed.

Monday, November 17, I called Peter to pick up and deliver the whirlpool tub for the master bath, Marc to set a firm date to install the fireplace and left a voice mail message for Naturally Stoned Chad to schedule the installation of the counter tops. Chad did not return my call, but by that time, I was conditioned to expect it. Shawn and the Top Dollar crew started installation of the cabinets. Shortly after he started the layout of the base cabinets, however, Shawn advised me that the cabinets would cover two receptacles and one wire for an overhead light if they weren't moved. Later in the day I called Brian and Jamie to relocate the misplaced wires.

CABINET PROBLEMS—Layouts for factory-fabricated cabinets are always redrawn by the cabinet sub-contractor, based on as-built (existing) dimensions. My kitchen plan was redrawn specifying a six foot long by two foot wide island base cabinet and a dish washer base cabinet flush with the end of the left, side wall in the kitchen. Although Top Dollar prepared the redrawn layout, it nonetheless ordered and started installation of a five-foot island base, and a dishwasher base one foot from the edge of the wall. In addition, it ordered and installed two base cabinets in the master bath that were also shorter than the space allotted.

Situation—My plan specified dimensional requirements for all base cabinets and the cabinet subcontractor verified the as-built layout. My contract price was based on the specified and verified dimensions. Kitchen and bathroom cabinets were ordered and being installed with shorter lengths. If you were the general contractor, what would you have done?

My Decision—Stop installation of the cabinets until all of the required sections were received.

DOOR PROBLEMS—Later Monday afternoon, the interior doors were delivered. I inspected the doors for damage and compared the number delivered, their sizes and directions of swing as shown on the delivery ticket against my original order. The inspection revealed six right hand doors were missing, three additional left hand doors were delivered and molding was broken on four of the jambs. My original order showed the doors as they were ordered. The delivery ticket, however, showed that all doors on the original order were delivered with none back ordered. As far as the supplier was concerned, the order was correct and complete.

Tuesday morning, I called my sales representative to discuss my door order and learned that he was out of town for the rest of week. I spoke with the store manager and he informed me that he would check the order and call me back. Within the hour, he called back and said he ordered the six right hand doors. He said he placed the order on rush and would try to have them on my project by Friday morning. Peter called a short time later and advised me that the tub for the master bathroom had not yet arrived. I called Hot Stuff to schedule the fireplace installation. As usual, I had to leave a message on Marc's voicemail. As usual, Marc did not return my call. Chad called, however, and informed me that the granite for my countertops was delivered to his shop.

While I was painting that following Sunday, Joan did a walk through and noticed the receptacle specified for the left side of her lavatory in the master bathroom was not installed. She also noticed that the cabinet above the refrigerator and the appliance garage did not have the correct doors.

Monday, November 24, I contacted Shawn and told him to add the correct doors to the cabinet order. I also called my sales representative and inquired about my door units. The representative informed me that the doors left the yard that morning on the first truck. Later in the morning I drove to the project to meet with my tile subcontractor. When I got there, Kenny informed me that the doors had not yet arrived. Thinking that the doors were in route, I called the dispatcher at the material supply store to get an estimated time for delivery. The dispatcher informed me that my doors had just arrived at the woodshop and needed the trim installed before they could be sent to my house. I informed him that I had a crew on site and they needed the doors ASAP. He said he would put a rush on the trim work and have the doors on my project before noon the next day. I advised the dispatcher that his time line was not acceptable. At that point, I called my sales representative and informed him of my situation. He told me the doors would be on site early Tuesday morning.

After I met with Joe "The Tile Guy," a member of the Good Brothers crew told me that the urinal inside the portable toilet was broken. I called the toilet rental company and requested a replacement. My trip to the job cost me another two hours and 45 minutes of annual leave.

Tuesday morning, the doors were not on site. Four calls to the supplier, from 8:00 AM to 3:30 PM, resulted in four separate estimated times the doors would be delivered. I arrived on the jobsite at 5:30 PM and was informed by Kenny that the doors arrived, but every one was hinged on the opposite side, therefore, he rejected the order.

Situation—Two orders of interior doors were rejected because of improper sizing or milling. The doors were the next components to be installed. Without them I would be loosing valuable time. If you were the general contractor, what would you have done?

My Decision—I called the general manager of the material supply store, stressed my immediate need for the doors and demanded that they be on my jobsite by 8:00 AM the following morning. They were.

Tuesday evening, I called The Tile Guy and was informed by Joe that he did not have the time to work on my bid, but he would have it ready by the next day. I also called Hot Stuff and left a message for Marc to return my call. Wednesday, no calls and very little progress was made for the balance of the week.

Monday, December 1, Marc finally called and said the fireplace would be installed the following day. Hot Stuff showed up late Tuesday afternoon, but did not have enough flue pipe and could not complete the installation. As the crew was finishing for the day, the crew chief asked me where the wire for the fireplace blower motor was. It was not there. I called Brian and asked him to install the wire as soon as he could. That afternoon I scheduled the installation of the attic insulation for next day. Wednesday, Brian installed the blower motor wire and the attic insulation was installed.

Whenever I have blown insulation installed in the attic, I prefer to have it done before the soffits and attic vents are in place. That sequence allows the over blown insulation to fall through the open (soffit) area below the roof overhang, preventing the vent holes in the finished soffit from being blocked. The soonest "Vinyl Bob," my vinyl subcontractor could install the soffits was the following Monday.

THE ILL WIND—On Friday, I took two hours of annual leave and met with Joe on the jobsite to go over the tile work. Chad's crew worked on the counter tops until it got dark, but did not finish. Chad assured me that his crew would finish Saturday. That did not happen.

Late Saturday afternoon, an unsuspected and unusually strong wind developed. Without the soffits in place, the open area below the overhang allowed the wind to blow in one end of the attic and out the other, drawing pink, pea sized pieces of insulation with it. When the wind finally subsided, my neighbors yard looked like it was covered with pink snow. Fortunately my neighbor and his family were away for the weekend. I spent most of Sunday raking and picking my blown insulation particles from my neighbor's grass and cleaning mud from the street.

ODDS AND ENDS—Monday, December 8, I went to the house before work to meet with Peter Plumber. Pete forgot our appointment. I also unlocked the front door so tile, cabinet and counter top work could be performed. All three forgot their commitments to work as well. Vinyl Bob was scheduled to install my soffits. He forgot also. Frank, with Life's a Gas showed for his appointment to discuss a revised bid, however, which salvaged the morning. Meanwhile, I left word on Danny's voice mail to call me and give a price to grade my driveway and haul construction trash. Danny was obviously busy. He did not return my call. The morning cost me another three and a half hours of annual leave, with very little to show for it.

Later in the day I called all of my "no shows" and informed them that I expected their crews on my job the following morning. Tuesday morning, I went to the house before work to unlock the front door. Peter Plumber, The Tile Guy, Top Dollar and Vinyl Bob showed up for work. Naturally Stoned Chad did not. When Shawn arrived, he promptly informed me that another receptacle had to be relocated before he could go much further. Knowing that Brian would need at least a day's notice, I relocated the receptacle myself.

Wednesday still no Chad, revised gas bid or grading bid. Thursday I took day of annual leave to be on job. I met with Efrain and Joe to discuss what was necessary to complete their tasks. Efrain said he would have several workers on the job Saturday morning to finish the brickwork. As I worked in the attic, I noticed several areas where the blown insulation was not installed to the appropriate depth. I called Craig at Long John's and informed him of the areas that needed additional insulation. Saturday I picked up glass block and mortar for the shower enclosure in the master bathroom.

Monday, December 15, I called Lumpy Larry to schedule the drywall walk through inspection, Danny to get my grading bid and

Craig to schedule a date to finish the insulation. Danny said the bid was at his home and that he would call me later. That did not happen. No one answered the telephone at Long John's. Tuesday I called Shawn and he promised to be on the job the following Thursday. I also called Frank at Life's a Gas to get my revised bid. Later in the day, Marc called me to arrange for payment for the fireplace. We agreed to meet on the job the following Thursday.

Saving all receipts for payment of services or materials is critical. Late Tuesday afternoon, I got a telephone call from the subcontractor I used to cut my concrete floor slab, requesting payment. I advised the caller that the invoice was paid by credit card the month before. That evening, I located the receipt, which was marked paid, and faxed it to the company. The issue was closed. Also that evening I called Danny and Craig again and left word to return my calls.

I purchased a granite kitchen sink to be inlaid into the granite counter. The sink could have been mounted above or below the counter top. I wanted the sink mounted above the counter top and made that desire perfectly clear to Naturally Stoned Chad. Wednesday the counter tops arrived and were installed, along with the kitchen sink before I got to the project. You guessed it. My sink was installed below the counter.

Situation—My subcontractor received explicit instructions to mount my kitchen sink above the counter top. The sink was installed below the counter top instead. Although the sink was mounted and perfectly useable below the counter top, it was not in accordance with the agreement I had with my subcontractor. If you were the general contractor, what would you have done?

My Decision—I had my subcontractor remove and remount the sink.

Situation—All but a small section of the ceramic tile had been laid and I still had work to be performed inside the house by other subcontractors. If you were the general contractor, what would you have done?

My Decision—As the contractor, it was my responsibility to protect all of the finished components and systems from damage while the house was still under construction. For all finished surfaces that could be stepped on, stood on or used as impromptu workbenches, I cut and taped down thick sheets of cardboard. For the bathtubs, shower units and other open spaces, I covered the open areas with six-mil polyethylene and taped the edges.

Thursday I took a day of annual leave and scheduled meetings on site with Marc, Craig, Lumpy Larry, Joe, Frank, Shawn, Chad and Tim, the owner of "Hard Time Concrete," the concrete finisher for my driveway. Craig, Larry and Joe completed their work, Shawn's crew worked, but did not finish and Marc, Frank, Chad and Tim did not show up. No work other than painting, which I was performing, occurred from December 19 to 30.

THE END IS ALMOST NEAR—Tuesday December 30, the kitchen island was set and the exterior sections of the gas lines were installed. I hired "Yardbird Landscaping" to perform most of my landscaping. The first landscaping project was to build a short retaining wall to raise the grade and hold back the earth for a parking area adjacent to my garage. As usual, before work was started, I met with Rickey, the owner of Yardbird and informed him that I wanted landscaping fabric installed on the inside of the wall to keep the fill dirt from sifting around the sides of the retaining wall blocks. Rickey built the wall himself, but forgot to install the fabric.

Situation—Whenever I design retaining walls, I specify landscaping fabric to allow water to move through the wall without taking fine particles of earth with it. Doing so prevents the soil from sifting and erosion, thus keeping the retained earth intact. My landscaping subcontractor installed my retaining wall, filled and compacted the earth, but forgot to install the fabric. If you were the general contractor, what would you have done?

My Decision—Remove all dirt five feet in from the retaining wall, install the required fabric, refill and compact the earth.

Wednesday, the garage doors were delivered. I got to the house as the installer finished hanging the first door. My garage doors were ordered with glass in the top panels to allow more natural light into the garage. When I questioned the installer about absence of the glass panels, he called his dispatcher to verify my order. The dispatcher verified that the wrong panels were sent to my job. I liked the additional security and insulation value of the solid panels, but was concerned about the amount of natural light that the one wall window would provide. The window, however, was rather large for a garage so I told the installer to install the second door. We agreed that I would check the light level in the garage over the next several days and, if I was satisfied, would keep the solid panels. The light was fine; I accepted a price reduction and kept the solid panels. Tim with Hard Time Concrete called and scheduled the formwork for the driveway to be installed on the following Friday.

Friday, Hard Time started the grading and formwork for the driveway. I like the concrete for my driveways and parking areas to be a minimum of six inches thick, therefore, the amount of concrete required would take two days to place. The first pour was scheduled for the following Monday.

Monday, January 5, the temperature was forecasted to drop to a low of 26 degrees Tuesday morning and 18 degrees on Wednesday morning.

Situation—Concrete should not be placed without sufficient precautions to protect against freezing when the air temperature drops below 40 degrees within a 24 hour time period. The temperature was forecasted to drop well below freezing for two straight days after the scheduled placement. If you were the general contractor, what would you have done?

My Decision—I called Tim and told him to stop the delivery of the concrete and reschedule it for the following Thursday.

Wednesday, I took two hours of annual leave to accept delivery of my major appliances. Thursday, I had to ask Tim to cancel delivery of the concrete again due to rain.

My kitchen range hood was custom made from cherry wood and required to have a stainless steel interior liner. Under normal conditions, the mechanical subcontractor installs the finished range hood. Since Top Dollar installed the wood portion of the hood, Gene declined my request to install the steel liner. After several telephone calls Thursday morning, it was obvious that I would not locate anyone that would be willing to install the hood liner. That left no alternative than to install it myself. Due to soggy soil conditions, I also called Tim and asked him to rescheduled the concrete for the following Monday.

Hood liners are manufactured with specific dimensions. From the exterior dimensions of the steel liner the interior dimensions of the wooden hood are established. The hood is built slightly larger than the liner to create a good fit. My liner dimensions were given to, and verified by Top Dollar when the hood was ordered. When I attempted to mount the liner, however, I discovered that the hood was 3/16 of an inch too short. On Friday I called Shawn and advised him about the size difference between the hood and its liner. He said he would be on job 10:00 AM the next day. I agreed to meet Shawn early Saturday morning, then headed to the lighting supply store to pick up some of the light fixtures.

Saturday Shawn showed up on time and modified the hood to accept the liner. With the assistance of Joan, I installed the liner, the motor, the blower unit and the flue piping. Sunday Joan and I shopped for more light fixtures and ceiling fans. I purchased a utility sink for the garage and backyard bathrooms and delivered them to the house.

Monday, January 12, I took one hour of annual leave to unlock the door for Brian and go over the light fixture placement. Joe was scheduled to work, but did not show up. Tim's crew showed up at 10:00 AM and Rickey a half hour later. I took an additional half-day of annual leave to be on site to supervise the concrete placement and landscaping work. When I got to the job, Rickey informed me that he cut the underground telephone line running to my neighbor's house. I called the telephone company and ordered an emergency repair.

The concrete was delivered late and Tim could only place and finish half of the planned area. Brian showed up in the afternoon and started installing the light fixtures. He later informed me that the foyer light fixture was defective and missing parts. I contacted the light fixture supplier and ordered a replacement. The telephone company failed to show, so I waited on site until my neighbors got home (8:30 PM) to explain to them why their telephone service was out of order.

Tuesday Peter started setting the plumbing fixtures. Brian and Joe did not show up. Tim and his crew placed and finished the concrete they could not complete the day before. When I performed my walk-through later in the day, I noticed three of the water closets that Peter set were too short in height.

Situation—I ordered and paid more for water closets with higher seat heights (referred to as "comfort height" or "chair height" water closets). My plumbing subcontractor picked up, installed and connected standard height water closets that were provided by my plumbing fixture supplier based on my previous order. If you were the general contractor, what would you have done?

My Decision—Remove the standard water closets and replace them with comfort heights. I also called my plumbing subcontractor and advised him to bill the fixture supplier for the time required to make the corrections.

Wednesday I took a day of annual leave, scheduled the power company to dig the trench and install the service cable for my underground electrical service and lined up Joe to finish grouting the tile. Lotta Air was scheduled to set my HVAC condensing units as soon as the power company closed the trench. Late morning the gas inspection was performed and Life's a Gas given permission to close the gas line trench. Marc finally showed up to be paid for the fireplace. Early afternoon I called Brian. He showed later that day and set several of my major appliances. Shortly before sundown, the field engineer for the power company showed up and informed me that the trenching crew was behind schedule and my work had to be rescheduled for the next day.

Thursday Tim and his crew showed up, but the dispatcher at the concrete plant said several drivers were out and he could not schedule delivery until Friday. The power company did not show, so I called Gene at Lotta Air and informed him of the delay. While tamping the fill for the gas line trench, I noticed the gage for the gas line pressure test was reading zero.

Situation—My gas line was pressure tested and held the required pressure for one day prior to inspection. The inspection was performed and approved. The following day, the pressure gage was reading zero. The pressure could have been relived by several possible causes including intentional "bleed down." If you were the general contractor, what would you have done?

My Decision—I called Frank and asked if a member of his crew performed a bleed down to relieve the test pressure. When Frank said "no," I asked him to retest the entire system for a leak.

Friday the power company did not show up. Tim, however, finished the concrete.

Monday, January 19, Peter's crew worked. As they were preparing to leave, I noticed the crew failed to set the dishwasher and the water closet in the backyard bathroom. I confronted the crew chief and informed him that I wanted him to complete his work before he dismissed his crew. He did. Later that evening, I reviewed the invoice for my plumbing fixtures (picked up and paid for by Peter) and noticed that the plumbing fixture supplier over billed Peter in the amount of $1,030.58.

Tuesday Joe grouted tile, Shawn worked on the bathroom cabinets and Brian hung light fixtures. The power company still could not find my address and failed to show again. When I got to my office, I called Peter and the plumbing fixture supplier and pointed out the invoice discrepancies, advised them to check their records and requested they call me back. In addition, I called the power company and asked for a firm date for the trench and service cable installation. The construction manager for the power company assured me that a crew would be on my job the following morning. I called Gene and put him on notice. Frank's crew retested the gas lines and verified that there were no leaks. Later that afternoon, I picked up more light fixtures and delivered them to the house.

Thursday I took a half-day of annual leave to meet Brian on the job and to be there when the power company arrived. The crew from the power company showed up mid morning. Shortly after the crew started trenching within the utility easement, the trencher cut the main television service cable. I placed an "emergency service" call to the television service provider and scheduled a repair crew for later in the day. The power company's crew finished its work, but could not fill in the trench until the television cable was repaired. Shawn worked on the bathroom cabinets. Brian showed up at 5:00 PM but did not work. Instead he promised to have full crew on the job Friday morning. The television and power companies finished after sunset. It was dark and cold so I called it a day.

Friday Brian's crew arrived early and worked about three hours. I got to the house after the crew left. When I checked the work, I found one exterior flood light mounted in the wrong place, the junction box specified to serve the pool equipment not installed and the interior garage lights installed perpendicular to the garage doors rather than parallel as specified on the plans. When I returned to my office, I called Brian to inform him of the discrepancies and Joe to schedule the completion of the tile work for the following Monday.

ONE MORE STORM—Sunday an ice storm hit Columbia late in the morning. The weather forecast called for low temperatures near freezing for the next two days. As much as I hated to do it, I had to call Joe and delay his work.

Monday, January 26, until the following Friday, no work occurred due to the cold temperatures. Saturday Peter installed a water line to the back yard for the swimming pool.

NOTE! When performing interior finish work, it is essential to maintain an even and consistent temperature in the house at all times. Materials such as tile grout, wood flooring, carpet, wood trim (especially at mitered joints) are affected by temperature and humidity. You could use portable heaters during working hours for the workers comfort, but to properly acclimate the materials being installed and to reduce the risk of immediate or future damage, expansion or shrinkage due to extreme day and night temperature differences, you should heat the entire house.

THE FINALS—Monday, February 2, I called for my final electrical inspection to allow for the electric meter to be installed. When I requested the inspection, I asked to be called in advance so I could be on the job to unlock the house before the inspector arrived. To follow up, I left the same request on the inspector's voice mail.

Tuesday there was no call from inspection department. When I went to the house later in the day, an inspection ticket was hanging on the front door stating that the inspection was not conducted because the house was locked. I called the inspector and left word on his voice mail for him to call me between 7:30 and 8:00 the next morning to reschedule the inspection. Wednesday still no return call. I called a total of seven more times until I got a live person on the line. The individual informed me that the phone lines were down due to office modifications. I rescheduled the inspection for Thursday morning and asked for a call before the inspector got to the house, and followed up again with a voice mail on the inspector's telephone.

Thursday I did not receive a call from any member of the building inspection department. On a hunch that the inspector would show up in the afternoon, I went to the house during my lunchtime and unlocked the door. I went back to work and called the inspector just before quitting time. He informed me that he passed the inspection and authorized the electric meter to be installed. With heat in the house, I was able to schedule completion of the remaining finish work that I was not going to perform myself.

Monday, February 9, I took five hours of annual leave. By the end of the day, the house was ready for the final inspection.

Tuesday I called the building inspection department and made an appointment with the inspector for 9:00 AM the following day for the final inspection. The house passed. All remaining work (additional trim, paint, interior door hardware) was under my control. With heat and fully functioning bathrooms, I could work inside at my own pace. The end was in sight.

WRAPPING UP—As I got closer to our move in date, I realized how long and exhausting this particular project was. Every time I stopped work on my project or required a subcontractor to remove and reinstall a component I added time to the project. That time cost me money, especially after I made the first draw on my construction loan. In all fairness I will say the anomalies encountered on this project were more than I had encountered on other construction projects. I will also state, however, that other projects that I am aware of had more problems.

WHY I DID IT—The reason I presented my case history in painful detail is to illustrate my (and your potential) personal time and involvement in building a house as an owner/contractor. This chapter was not written to discourage you, but to prepare you for the realities of a major do-it-yourself construction project. If you believe that you can run a construction project by trusting your subcontractors to do what they promise; without direct hands-on involvement on a daily basis; without making critical decisions; without confrontations; without making many telephone calls; without disappointment; you are setting yourself up for failure. Construction projects present many challenges and many highs and lows. Things happen fast and decisions that affect the cost or outcome of the project must be made just as fast. You cannot sit in your home or office and expect your house to build itself.

Whether it was by irony, fate or absolute coincidence, Joan and I moved into our new house on April Fool's Day.

CHAPTER TWELVE

CONCLUSION

My motivation and goal in writing this book was to save its readers money by pointing out the benefits of assuming the role of owner/contractor in home construction projects. The book was not written to convince you that you could succeed as a contractor or even encourage you to try. Those are decisions that only you can make. My intention was to lay out the process in a logical order, present an objective view of the risks and rewards and give you the necessary knowledge to make your own decision.

FOOD FOR THOUGHT—During the course of the project highlighted in my case history, I hired 39 different subcontractors and purchased materials from 19 different material suppliers. To get the subcontractors I hired, I contacted dozens more to request bids. Some were no shows, some would show up and look at work to be performed but not provide a bid or even call back. Each subcontractor and supplier required my time and attention. In most instances, as you may have noticed, some of my subcontractors required much more of my personal time and attention than others.

Many subcontractors will cut corners, short you on materials or cover mistakes if possible. As the contractor, you are responsible for inspecting the installation of the materials and components for all systems and the workmanship of all subcontractors, identifying the anomalies and requiring their correction. Unless the construction industry is very slow at the time you build, over booking and scheduling problems will most likely happen. Remember, as an owner/contractor, you will be at or near the bottom of your subcontractor's and material supplier's priority lists. Lying on the part of subcontractors and material suppliers is common and should be expected. If it does not happen to you, consider it a bonus.

Reliability, quality, experience and pride in workmanship on the part of subcontractors in all construction trades is declining. It is very important that you devote ample time to verify the claims and credentials of all subcontractors that you consider and, if possible, inspect their work and definitely check their references prior to hiring. If the construction industry is busy at the time you build, you may not get the subcontractors that you want and may have to wait for the ones you settle for.

As an owner/contractor, there are things you can control and things you cannot control. You can control who works on your job, the quality and workmanship of your subcontractors (based on reputation), the materials used and the overall cost of the project. You cannot control the weather, delivery of materials or equipment, the actions, honesty, attitudes or habits of your subcontractors. You must be willing to thoroughly scrutinize the work of each subcontractor and be prepared to take a position on the quality of its work, stick to that position and deal with uncomfortable confrontations as they arise.

Before you start a project, you should have at lease a basic knowledge in reading construction plans and have the ability to identify common construction materials. If you cannot distinguish a joist from a rafter, a truss from a header or at any time have to refer to any component as a thingie, you should take a little more time to familiarize yourself with basic construction materials. In addition, it is essential that you have the time to inspect and the ability to spot discrepancies after each subcontractor completes every phase of its work.

No matter how explicit your instructions are, they may not be thoroughly understood by the workers that perform the actual tasks. Do not assume that your thoughts, words or intentions are creating the same picture in your subcontractor's mind that you have in yours. You must clearly state what you expect and be assured that your subcontractors understand your instructions. It is up to you to take a firm position and insist or even demand that your instructions are met to your satisfaction. As the contractor you need to be familiar with the tasks that need to be performed and where the various components need to be installed.

READY FOR A DECISION?—By now you should have the knowledge to make a sensible decision about starting a construction project of your own. You also have the knowledge to save money. The only question remaining for you to answer is how that money will be saved. Will your money be saved passively by not pouring it into a project that you do not have the time, stamina or motivation to complete or will it be saved actively by assuming the role of the contractor? Either way is ok. Either way you win.

Let's revisit the four groups that I mentioned in the Preface and see where you fit in.

Thinkers—If you still have not come to a solid conclusion about your ability to succeed as an owner/contractor, you may have none-the-less answered the question without realizing it. If it is your nature to think of every possible scenario before making a decision, my advice to you is, do not attempt a project as an owner/contractor. The activities on a construction project require decisions to be made and quick action to be taken on a daily basis. Thinking too long about a construction problem, an alternate method or material, selection of a subcontractor, etc., will drive you crazy and result in loss of time and money. Put the book away, hire a general contractor to perform the work and realize that you have passively saved money by not starting a project that may take an indefinite time to complete.

Fence Sitters—At this point, if you need to be coaxed or otherwise assured that you have what it takes to succeed as an owner/contractor, your comfort level is still too low. If you feel the need to have someone with construction knowledge assist or advise you and cannot freely jump off the fence and assume full responsibility, my advice to you is, do not attempt a project as an owner/contractor. To succeed in construction, it is critical that you have complete confidence in your ability to manage the project from start to finish. Hire a construction manager or a general contractor to perform the work and realize that you have passively saved money by not starting a project that may stall or stop altogether due to lack of confidence on your part.

Hardheads—Soften up! You do not know everything. Don't pretend that you do. If your decision is to go ahead with your project, my advice to you is to read and understand everything I have offered in this book. When you run in to an obstacle, ask for advice from those that have the answers, and then heed the advice given. You will be risking your own hard earned cash. Don't be foolish with it. If you feel that it is not possible for you to follow that simple advice, hire a general contractor to perform the work and realize that you have passively saved a lot of money by not running your project in several different directions before you run it into the ground.

Doers—If you started as a doer or moved from thinker, fence sitter or hardhead to a doer, you are in position to move forward and succeed with your construction project. Take the time to learn all the necessary processes, ask questions, stay informed and devote the time required to keep the project running smooth. Know all tasks in advance and perform them in a logical manner and not only will you succeed, you will have great satisfaction and actively save a fair amount of money.

ONE LAST THOUGHT—As in all endeavors that require a degree of skill, experience is always the best teacher. No book, regardless of how well written can provide, substitute for or replace that very valuable element. If you have the time, consider working on a construction project before you start your own. By doing so, you can get a feel for what you will doing on a larger and more personal scale, gain some valuable experience and become familiar with the overall construction environment. Go into your project as prepared as possible and then do whatever is necessary to complete it.

GOOD LUCK!

INDEX

N

Neighborhood 21, 34, 51, 65, 73, 79, 101
    conditions 37
    convenience 34
    existing homes 37
    growth 16
    mail 19
    market 21
    standards 13, 65
    tenant/occupant mix 18, 34
    theft 164
    transactions 21
    use 15
    utilities 121
    zoning 15, 16, 18

O

Occupancy 148, 149, 168, 169, 171
    certificate of 148, 149, 168 - 170, 172
Overbuild 24, 37
Overpayment 150

P

Payment 27, 74, 77, 80, 82, 85, 102, 105, 113, 114, 157
    advance 80, 85, 86
    cash 78
    down 27, 85
    fees 18, 67, 76, 121
    final 105
    history 73

    internet sales 159
    late 73
    method of 105, 110, 113, 114
    request for 113
    requirement for 114
    taxes 80
    withholding 114
Permit 52, 67, 111
    application 122, 123, 125
    building 37, 52, 65, 69, 75, 119, 121, 123 - 126, 128
    burning 133
    checklist 126, 133
    construction 15, 69, 79
    electrical 52, 69, 119, 121, 126
    encroachment 119, 121, 126
    excavating 69, 126
    fee 52, 67, 111, 125
    gas 52, 69, 119, 121, 126
    grading 119, 121, 126
    issued 39, 65, 121, 123, 126, 128, 134
    mechanical 52, 69, 119, 121, 126
    obtaining 44, 76, 112, 119
    placard 103, 128, 165
    plumbing 52, 69, 119, 121, 126
    posting of 128, 134
    preconstruction 125
    processing 123
    receipt 123
    required 52, 119, 125
    seawall/bulkhead 119, 121, 126
    septic tank 119, 121, 126

R

Z